IN ROBOTS WE TRUST

SAMUELE VINANZI

IN ROBOTS WE TRUST

OXFORD
UNIVERSITY PRESS

OXFORD
UNIVERSITY PRESS

Great Clarendon Street, Oxford, OX2 6DP,
United Kingdom

Oxford University Press is a department of the University of Oxford.
It furthers the University's objective of excellence in research, scholarship,
and education by publishing worldwide. Oxford is a registered trade mark of
Oxford University Press in the UK and in certain other countries

Published in the United States of America by Oxford University Press
198 Madison Avenue, New York, NY 10016, United States of America

British Library Cataloguing in Publication Data
Data available

Library of Congress Control Number: 2025932100

ISBN 9780198952916

DOI: 10.1093/9780198952947.001.0001

Printed and bound by
CPI Group (UK) Ltd, Croydon, CR0 4YY

Links to third party websites are provided by Oxford in good faith and
for information only. Oxford disclaims any responsibility for the materials
contained in any third party website referenced in this work.

The manufacturer's authorised representative in the EU for product safety is
Oxford University Press España S.A., Parque Empresarial San Fernando de Henares,
Avenida de Castilla, 2 – 28830 Madrid (www.oup.es/en
or product.safety@oup.com). OUP España S.A. also acts as importer into
Spain of products made by the manufacturer.

To my wife Crefeda, who taught me more about trust than any robot ever could.

CONTENTS

FOREWORD

In Humans we Trust?
Dr Scott Midson

After having read Vinanzi's insightful and illuminating book about the importance of trust for how we exist in the world, I find myself taking stock of all the mundane and otherwise overlooked things that I necessarily trust to go about my daily life. Of course, I trust the people involved in all the social relationships – friends and family – that have a significant impact on my identity and on my psychological and social well-being. Beyond that, though, as Vinanzi encourages us to consider, there's a whole network of other relations with not only people I don't know, but also with non-human things, without which I couldn't live my life. As he writes, "at a very basic level, we do trust or distrust the objects around us, but in a much different way than we would trust or distrust a person."[1]

I trust, for example, that when I pay for goods or services, they are of a standard as advertised; here, I trust the people in various roles that work to deliver any kind of product, as well as also having trust in the quality of the product itself. If I were to buy a free-range egg, for example, I would place trust in the farmer, but also in the various agents involved in checking and ensuring

---

[1] p. 47.

standards and fair prices, as well as in those involved in marketing and selling the product to me. I also trust the egg to not harm
me when I cook it – this is a seemingly different (but, as Vinanzi
says, a no less significant) type of trust to trust in other people,
but essentially, I'm trusting the wisdom of experts who work to
ensure the safety of the egg (including instructions for how I can
safely cook it, in which case I'm placing trust in my own ability
to follow those instructions successfully). When it comes to having trust in other products, again I note all sorts of entanglements
and things that I must place trust in. To provide another example
here: I trust that my house will do what I expect of it – namely, that
it will keep me and my family (and my belongings) safe and dry,
and not collapse. (This admittedly isn't always an easy act of trust
when living in an old house and a new crack appears on a wall or
ceiling.) In that, I implicitly trust the people who built my house
and the work of everyone since – previous owners and tradespeo-
ple – to upkeep it. I often also find myself, when considering my
own role in upkeeping the house, trusting the wisdom of various
people who share their building tips and tricks on the Internet,
which is perhaps not always the least questionable of things to
place my trust in. When evaluating different comments and suggestions online, though, I'm also required to trust myself in my
own judgments and deciding a course of action.

Across these examples of where I place my trust in the things
around me, I find that it is generally a trust that is rooted in
received wisdom – i.e. in people who have designed different
things, and later in people (including myself) who produce
things according to those designs. In other words, I trust the
knowledge and capabilities of people to make things that themselves can become dependable and trustworthy. But wisdom

is more than just knowledge: Robert J. Sternberg explains how wisdom is different from intelligence in that it is about the application of intelligence, as well as knowledge and creativity, towards a *common good*.[2] This point links to what Vinanzi discusses of Aristotle's notion of *'phronesis'* as practical wisdom rather than "cold, analytical thinking,"[3] but it puts it more explicitly in terms of our connections to, and indeed our relationships with, others in society. The idea of the "common good" has been discussed by many groups from philosophers to politicians, but it's essentially about working to support the flourishing and betterment of people that you will likely personally never meet; it's therefore about an indirect relationship with others who are impacted by your work. By benefitting from the things around me, and by recognizing the wisdom of people who put those things in place, I effectively express trust in others working towards the common good.

In addition to the things – the people, systems, and services – that I trust, there are all the countless agents that trust me. To name a few: my employer and my students expect me to teach to the best of my abilities (or at least to what they might come to expect that standard to be); my dog trusts me to take him for walks and to ensure he's fed (usually at routinized times) and watered; my mortgage provider trusts me to keep up with my monthly payments. That I am trusted to fulfill duties and meet expectations of these people (yes, I include my dog in that claim of personhood!) suggests that I am *accountable* to them.

---

[2] Robert J. Sternberg, 'Race to Samarra: The Critical Importance of Wisdom in the World Today', in Robert J. Sternberg and Judith Glück (eds.), *The Cambridge Handbook of Wisdom* (Cambridge: CUP, 2019), p. 5.
[3] p. 86.

My accountability suggests that there is a degree of trust placed in me to act, as Vinanzi astutely puts it, *reliably* but also in a way that reflects my *commitment* to the person that I am acting in relation to – and indeed that I am in relationship with.[4]

To the extent that I am accountable to these people that trust me in various ways, I can speak of my *responsibilities* to act in a certain way in relation to them. "Responsibility" is another tricky word to explain, as it can indicate taking responsibility *for* something or someone – i.e. as a dog owner, I am expected to take responsibility for my dog's actions in public, and as my dog (like nearly all dogs) isn't always predictable or indeed trustworthy, I must keep him on a lead, especially when around roads or busy areas. Here, I act in the best interests of my dog, but also in the interests of others (so that, for example, no one has to swerve to avoid an easily distracted dog running across a road, or no young child finds they have a miniature schnauzer unexpectedly barking in their face), in order to mitigate against risk. My responsibility here isn't necessarily about ownership or control (although that often comes into it in practice); in a strict sense, I am responsible for my own actions in relation to the welfare of my dog and others. This is philosopher of science Donna Haraway's point where she talks about "response-ability," namely seeing responsibility as acting in relation to others.[5] This definition of responsibility, as Haraway goes on to explain, isn't about ownership (*of* others) but rather accountability (*to* others) when acting in response to others. In this sense, there is response-ability between me and my dog, as well as between the

[4] p. 48.
[5] Donna Haraway, *When Species Meet* (Minneapolis: University of Minnesota Press, 2008), p. 71.

two of us as a relational unit and wider society (or at least those around us).

Accountability is an important way that we participate in notions of the "common good" that I introduced. Philosopher Amitai Etzioni makes this point in arguing that we should move from thinking about things like privacy as a "right," towards recognizing it as an "obligation": in other words, we should think not about individual aspirations or gains, but rather about how our actions respond to and impact others.[6] To that end, we hold ourselves accountable to generalized others – this, as Etzioni also points out, is crucial to the flourishing of society.[7]

All of this – trust, reliability, accountability, and the common good – is particularly interesting when it comes to thinking about robots. Do – indeed, *can* – robots have the same accountability as humans to the common good? Many in the newly emerged field of "*roboethics*" have debated this in asking whether, for example, robots can be programmed to follow ethical laws: US professor and sci-fi writer Isaac Asimov famously suggested a thought experiment to this end with his "three laws of robotics." (Although not many realize that there were actually four laws!) Asimov's laws were designed to demonstrate the limits of top-down, rule-based approaches to ethics (those "in the biz" call this *deontological ethics*), which we often think might readily lend themselves to the neat programming of robots. The issue, though, is that humans do not work in such neat ways: as many experiments have shown, so-called ethical laws often conflict one another or get compromised in different circumstances. Take the well-known "trolley problem," for example: if a trolley

---

[6] Amitai Etzioni, *The Common Good* (Cambridge: Polity, 2004), pp. 50–51.
[7] ibid., p. 2.

(or tram or train) is moving along a track towards two people, but you could switch its track to another where it would only hit one person, should you intervene? Is it better to save more lives? What about if the one person on the other track is a family member? Or a pet? The latter two points are particularly significant given what Vinanzi says about *commitment* being a key aspect of trusting relationships – we might say we have more commitment to those that we are closer to rather than strangers, which may impact our decision on whether to switch the track or not. As researchers in Harvard have shown, we can't even agree on how best to interpret ethical "rules" or guidelines, let alone expecting to be able to program it into a robot.[8]

A robot programmed to follow deontological ethics would likely not be able to manifest trust because, if you were one of the people on the track, and the robot was at the track-switching lever, you may not trust it to act in your interests if you can only speak of its reliability to follow programmed instruction rather than bank on its commitment to you. If Vinanzi is correct in how he figures trust as a key part of human–robot interactions, then the interaction is potentially limited given how the robot can build no accountabilities to users. On the other hand, other roboethicists such as Noel Sharkey have argued that we should not deceive ourselves as to what robots can and cannot do – to speak of a robot's "trust" or its commitment to a relationship or user may be to mislead ourselves as to the robot's sense of moral agency.[9] Vinanzi addresses the question of robots and moral

---

[8] Edmond Awad *et al.*, 'Moral Machine', *Max Planck Institute for Human Development*, https://www.moralmachine.net/ (2020).

[9] Noel Sharkey, 'Mama Mia it's Sophia: A Show Robot or a Dangerous Platform to Mislead?', *Forbes*, https://www.forbes.com/sites/noelsharkey/2018/11/17/mama-mia-its-sophia-a-show-robot-or-dangerous-platform-to-mislead/ (2018); cf. p. 19.

agency, concluding that, even with full transparency about a robot's functioning, we tend to perceive and subsequently treat them as moral agents.[10] This, he argues, is important for effective human–robot collaboration. Yet he also acknowledges what he calls "the pitfalls of undertrust and overtrust"[11] – what, then, is the "just right" amount of trust to place in the robot?

I wonder whether there are useful parallels that can be drawn here between robots on the one hand, and, on the other, the other non-human entities that I listed at the start of this reflection that I necessarily find myself placing my trust in, or that trust me. Firstly, in the case of my dog, I noted that he trusts me to feed him and provide him with water, walks, and safety. This is a testament to my reliability and my commitment to the relationship I have with him. My dog, however, like all dogs, is not always the most trustworthy of beings – in some capacities, I trust he won't destroy my house when I leave him alone for a few hours, but in other capacities, I don't trust he won't get distracted off a walk, which is why I walk him around most places on a lead. My sense of responsibility here mitigates against the sketchiness of trust, and it moreover helps me to think about *response* and *ability* between my dog and I, and additionally between the two of us as a relational unit and others. By this, I mean to show that there are multiple modes of response both within and beyond the relationship, and recognizing these can demonstrate what Vinanzi identifies as key markers of trust – a *commitment* to the other and/or to the relationship itself, and a degree of *reliability* – without necessarily having to demonstrate all of trust's connotations of moral agency. This is Haraway's point in talking

[10] pp. 52–56.
[11] p. 90.

about *response-ability*: she says that animals are "response-able in the same sense as people are; that is, responsibility is a relationship crafted in intra-action through which entities, subjects and objects, come into being."[12] It's a bit of a dense and abstract point, but basically, she's saying that responsibility is a way of recognizing meaningful relationships and interactions including with nonhumans. This doesn't nullify trust or its importance, of course, as responsibility (and accountability) and trust are clearly closely related, but it does offer a slightly different way of thinking about what it means to trust a robot, particularly in terms of the wider ethical significance of this.

It's also worth thinking about how robots, as well as having complex processing capabilities that make for seemingly meaningful interactions and relationships through modes of response and ability, are also created artifacts. To that end, they can be compared with other created things, such as some of the things that I listed at the start of this reflection that I place trust in, including the packaged and labelled "free-range" egg and my house. I suggested that these objects are effectively conduits for the trust I have in others who designed and made them, and in the received wisdom that brought them into being. Robots can be figured similarly. The trouble is, there is much distrust of technology companies including those that manufacture robots (and other AI): people have notable concerns about data privacy and misuse, as well as about manipulation, for example in drawing out emotional responses from users. Here, tensions emerge between money (i.e. from the sale of robots) and power

---

[12] Donna Haraway, *When Species Meet* (Minneapolis: University of Minnesota Press, 2008), p. 71.

(i.e. from the information contained within data) on the one hand, and things that generally come under the umbrella of "ethics" on the other (i.e. goodness, happiness, flourishing). Science fiction films frequently depict these tensions, such that, as Daniel Dinello observes, often the "bad guys" in films featuring killer robots are in fact the companies that make them (I'm looking at you, Skynet).[13] The argument from the perspective of the common good is that ethics must always outweigh personal or corporate gains, and this highlights peoples' responsibilities when designing, manufacturing, and marketing robots.

While work must clearly address the responsibilities of the people behind robots, as Vinanzi's work shows, it's no less important to consider the response-abilities of the people engaging with the robots, both directly and indirectly. I've highlighted this elsewhere in the context of Tay, an artificially intelligent (AI) Twitter-based chatbot produced by Microsoft that was designed to learn from users through interactions with them. Within 16 hours, Tay became homophobic, misogynistic, anti-Semitic, and racist after having been trolled by different Twitter users, and so Microsoft had to delete the account. In this case study, it's important to note the responsibilities of the designers as well as users, both of whom failed to act responsibly or ethically.[14] Although trust isn't directly suggested here, safeguards to mitigate against risk or potential untrustworthiness, or even to encourage the cultivation of trust, were not provided, which indicates a notable deficiency of responsibility. As humans who interact with robots

---

[13] Daniel Dinello, *Technophobia! Science Fiction Visions of Posthuman Technology* (Austin: University of Texas Press, 2005), pp. 131, 275.
[14] Scott Midson, 'From *Imago Dei* to Social Media: Computers, Companions, and Communities', in Scott Midson (ed.), *Love, Technology and Theology* (London: T&T Clark, 2020), pp. 146–60.

in various ways – including relationships *with* them as well as *through* them, i.e. to others and/or the common good – we cannot shirk our responsibilities.

What I'm effectively sketching out here in these reflections is the importance of considering the ethics of trust, by which I mean not just the question of whether or not it's ethical to make robots we find ourselves trusting, but that of how we figure trust in terms of responsibility (response-ability) and accountability, both to others in the context of specific relationships but also to the broader notion of the common good. Vinanzi's question, "can I trust a robot?" and its provocative and informative corollary, "can a robot trust me?" can be supplemented by questions about "to whom or what am I responding?" and "to whom or what am I accountable?" These are important questions that help us to reflect on our responsibilities, perhaps to robots but also *through* robots and to one another. To this end, we are compelled to ask not only whether in robots we trust but also whether in ourselves and in other humans we trust.

ACKNOWLEDGMENTS

This material is based upon work supported by the Air Force Office of Scientific Research, Air Force Materiel Command, USA. Funder award no. FA9550-19-1-7002. I extend my sincere thanks to them, especially to Nandini Iyer and Laura Steckman, for providing the necessary research funding and granting me the academic freedom to explore this fascinating journey at the intersection of humans and machines. Additionally, I would like to express my deep gratitude to my mentors, Antonio Chella, Angelo Cangelosi, and Alessandro Di Nuovo, whose continued support has been invaluable to my scholarly pursuits.

CHAPTER 1

INTELLIGENT MACHINES

A matter of trust

Robots have been a part of our collective imagination for decades, but now they are no longer just science fiction. In fact, they are becoming an integral part of our society: from self-driving cars to delivery drones, robots are coming to be an increasingly common presence in our daily lives. While witnessing their rapid advance, we are faced with the challenge of determining whether we can trust them or not. But what does it mean to trust a robot and why should we care? To understand the importance of the matter, let's consider the following example.

Imagine you've been invited to a job interview at a company you've never been to before. As you arrive, an employee at the reception hands you a visitor badge and offers to guide you to your destination. You accept and are led through a complex maze of corridors and stairs to a waiting room. Once there, the guide asks you to wait for your appointment and leaves you alone.

Time passes and you're still waiting, when suddenly a fire alarm goes off, piercing and urgent. Before you have a chance to react, gray smoke begins to seep into the room. Realizing the situation is escalating quickly and being all alone, you decide to leave the room and look for the nearest emergency exits.

As you step out of the waiting room, the scene is chaotic: the corridor is filled with thick smoke, your vision is impaired, and there's no sign of other people. You begin to search for the emergency exits, but the signs are nowhere to be seen. Suddenly, a moving light catches your attention. You approach it and discover a small mobile robot consisting of a metal base supported by two rows of wheels, equipped with a pair of glow sticks that it flaps like arms. Something is printed on its side: "Emergency Guide Robot." It becomes rapidly clear that this machine seems to be there exactly for that kind of emergency, possibly in place of the signage. The robot waves its luminous arms towards a specific direction, guiding you towards a short stretch of corridor that ends in a partially obstructed doorway beyond which lies a completely dark room. At this point, you're faced with a decision: do you follow the instructions of the robot, which seems to be performing the task it was designed for, or do you take your chances and try to find the nearest emergency exit on your own? Each alternative has its pros and cons, and your personal decision will boil down to one simple question: do you trust the robot or not?

It may come as a surprise that the previous anecdote is not just a story, but it is based on a scientific experiment that took place in 2016 at the Georgia Institute of Technology, in the United States of America [1]. The aim of the study was to measure the degree of trust that people are willing to place in a robot during an emergency. The results of this research are not only interesting but also unexpected, and we will examine them in detail in the following chapter (see Chapter 2). What's crucial to understand now, however, is that this example is a powerful illustration of why it's essential to start thinking seriously about trust in human–robot interaction: as intelligent machines become increasingly integrated into our lives, we will face a growing number of decisions that require us to rely on them. For instance, we may entrust a robot caregiver with the health of our loved ones, or a self-driving car to transport us safely.

Robots are complex machines that interact with us in a unique way, evoking a whole constellation of psychological, moral, and ethical reactions that have no equivalent in the artificial world. By understanding the factors that influence our behavior towards these machines, we can interact with them more deliberately and thoughtfully, gaining valuable insights into ourselves in the process. As robots become more common in our daily lives, these topics will become even more relevant. This book will help to deepen your understanding of technology and improve your computer literacy, enabling you to navigate your evolving relationship with machines. By exploring the complexities of human–robot interaction and the nuances of trust, we can prepare ourselves for a future where these machines will play an increasingly important role in our society. We will soon discover that our willingness to trust and our ability to judge the trustworthiness of others are critical components of our daily lives, with far-reaching and often invisible effects on how we interact with the technology that already surrounds us.

If you need more motivation to delve into the topic of human–robot interaction, consider the unprecedented technological progress we are experiencing today. Unlike in the past, when progress occurred gradually over centuries, nowadays we witness rapid, exponential advancements in technology and knowledge. In his 1999 book *The Age of Spiritual Machines* [2], futurist Ray Kurzweil proposed the Law of Accelerating Returns. This theory suggests that the rate of human progress increases over time, meaning that more advanced civilizations can achieve progress at a faster pace than less advanced ones. Kurzweil's law implies that this evolution is exponential rather than linear, which has significant implications for the trajectory of human development and the possibilities for future technological innovation. Our current rate of progress is not only setting new records but it's also expected to accelerate even further. Just think about how different our world is from the one our parents or grandparents were born into: in the span of just a century, our society has changed drastically. Smartphones, the Internet, virtual reality, social networks, green energy, and of course artificial intelligence are wonders that have been invented over the course of the last few decades, a very insignificant frame of time when compared to the length of human history. Innovations are becoming bigger and more frequent, leading us towards a very interesting future. However, rapid technological change can also have negative consequences, such as the need to adjust to the idea that our world will undergo drastic transformations many times during our lifetime. Acknowledging this should help us get into the habit of preparing ourselves for what lies ahead, to avoid being staggered by the arrival of disruptive technology.

We have made this mistake multiple times, particularly in our response to artificial intelligence. People and governments have been slow to react to the speed of progress, leaving us unprepared to face the socioeconomic consequences of these innovations. Let's consider a very recent case: the rise of ChatGPT. The latter is a software known as large language model: an artificial intelligence able to generate human-like text based on the input it receives. ChatGPT is capable of answering questions, writing stories and articles, generating poetry and songs, and even producing lines of code based on a prompt. The consequences of its release to the public in November 2022 have rippled through the fabric of our contemporary society. Those of us working in education had to suddenly face an enormous problem: from one day to another, students were delegating their essays and assignments to this software, in an unprecedented form of academic malpractice. Even students who don't cheat can sometimes face issues due to the use of artificial intelligence detectors to scan their submissions: far from being perfect, these systems can mistakenly accuse them of cheating. Similar problems are affecting the practice of journalism, where the ability of language models to generate text that is nearly indistinguishable from human writing has raised concerns about the potential for artificially generated content to replace humans and undermine the authenticity of news and information.

These examples highlight the need to consider the impact of new technologies on our lives before they become a reality. This is particularly true regarding the subject of this book: intelligent robots. Although they are not yet commonplace, the situation is rapidly evolving, and it won't be long before they become significant actors in our daily lives. To avoid being caught off guard by their arrival, we must learn more about them and prepare ourselves for the changes they will bring to our world.

The narrative in this book carries an important bias that I believe I should be transparent about, especially for readers who are not well-acquainted with robotics and are just beginning to explore this field. Due to my personal academic focus and the nature of my research interests, I am particularly interested in human-like robots. As we will see in Chapter 2, these are machines that have a body structure similar to ours, often displaying two legs, two arms, a torso, and a head, and they are primarily used to investigate the mechanisms of the human mind. The ultimate goal for myself and my fellow researchers in this field is to replicate the human mind in a machine, enabling a robot to perform all the kinds of tasks that we, as humans, can do.

From this perspective, the foundations of trust in human–robot interactions arise from symmetrical interests in the fields of psychology and engineering: the former seeks to understand the brain and the human mind, while the latter tries to build complex and intelligent machines that resonate with us and increase our well-being. These two disciplines are intertwined and mutually beneficial, with one providing insights into our mental processes and the other using this knowledge to build better artificial brains. This, in turn, can help us understand the nature of intelligence, creating a virtuous circle of knowledge exchange that benefits humanity. Throughout this book, we will encounter numerous examples of this interplay between disciplines, demonstrating how robotics can enhance a wide range of scientific fields.

On the other hand, this is not the only possible approach to robotics, and it would be enormously reductive to present it as a comprehensive representation of the field. Consider, for example, robot welders in the automotive industry or robots that are meant to inspect areas that are difficult for us to access, such as pipes, radioactive facilities, underwater cables, or even collapsed buildings. All these robots would be hindered by human-likeness, and instead need new bodies and specific forms of intelligence that would better serve their purposes. Even if we narrow our attention back to social interaction, which is the focus of this book, we can find examples where human-likeness is not required. Robot animal companions, for instance, do not require the nuances of the human mind to perform their functions. For these devices, animal-like intelligence is all that is required, and we will still be able to form relationships with them, including ones that involve trust.

Whatever the final objective, the pursuit of knowledge in the field of robotics is driven by a sincere curiosity and a desire to learn that are the hallmarks of the human spirit. As we continue to make discoveries in robotics, new philosophical, ethical, and moral issues have come to the forefront, especially as we approach a new kind of society in which we will be closely connected with intelligent machines. In this book, we will explore the nature of the relationship between humans and robots, taking care to avoid assuming any mathematical or technical knowledge on the part of the reader.

Let us embark on this journey by exploring the idea behind robots: their purpose and their current presence in our world.

Robots through the ages

The word "robot" was introduced for the first time by Czech writer and playwright Karel Čapek (1890–1938) for one of his most famous plays: *Rossumovi Univerzální Roboti*, translated into *Rossum's Universal Robots* and commonly abbreviated as *R.U.R.* [3]. Released in 1920, the play is a science fiction story about a company that mass produces artificial human beings using synthetic organic material, which are described as "lacking nothing but a soul." The play quickly gained influence, and in 1938 the BBC aired an adaptation, which became the first instance of science fiction on television (see Figure 1.1). Originally, Čapek had named these creatures *labori* after the Latin root of "labor," but in a later draft of his masterpiece he accepted a suggestion from his brother Josef who proposed to change the name into *robota*, a Slavic word, which means "forced labor." The latter had its origin in central Europe to indicate the compulsory labor of serfs within the feudal system (which was abolished in the 18th century by what English scholars call, curiously for us, the "Robot Patent").

Figure 1.1 A scene from the 1938 BBC adaptation of Karel Čapek's "Rossum's Universal Robots."

The origin of the name conveys the expectations that we have towards these machines: a robot is a piece of technology whose purpose is to take charge of heavy, repetitive, or unpleasant tasks on our behalf, freeing us from these burdens. In addition, we don't typically think of robots as conscious entities, but rather as servants that don't need to be compensated for their hard work: in other words, forced laborers.

What sets robots apart from generic machines is a degree of intelligence that allows them to transcend their status as mere objects and become more human-like. As a result, robots exist as hybrid entities, straddling the boundary between the inorganic and the organic worlds. While too intelligent to be considered mere objects, they lack the awareness and consciousness that are the landmarks of organic life. This powerful and revolutionary idea may seem like a modern development, but it actually dates back to the early days of civilization.

The concept of a mechanical, artificial human has been a recurring theme throughout human history, predating the modern scientific disciplines that now define the field. One of the earliest references can be found in Jewish mythology: the Old Testament refers to the *golem*, which in its etymological meaning refers to "embryo" or "raw material." According to the Kabbalah, an esoteric practitioner would have been able to create a body made of clay and bring it to life by speaking the correct sequence of magic words. The resulting entity would possess incredible strength, but would also lack in speech, wisdom, and judgment: essentially, an obedient servant to its creator.

As the wheel of time slowly turned, tales about automata continued to flourish. For instance, legend has it that in the fourth century BC the Greek philosopher and scientist Archytas, a pupil of Pythagoras, engineered a mechanical dove capable of flying from branch to branch. Throughout the Middle Ages, there were many stories of remarkable inventors who were said to have built machines capable of performing a wide range of tasks. In his book *Descartes and Medicine* [4], Gerrit Arie Lindeboom tells us the following story:

> Further, one should not forget that in the Middle Ages there was a vivid interest in automatons. Albertus Magnus (ca. 1200–80) had in his laboratory at Cologne a robot that could move and give a greeting saying "salve" (hail). There is a narrative that the inquisitive Thomas Aquinas once entered the laboratory during the night, but was so frightened by the unexpected welcome that he hit the robot so that it was broken into pieces.

In many of these cases, the distinction between myth and reality is blurred due to the lack of archaeological evidence: many of these prodigious machines seem to have shared a common fate of destruction that makes them impossible to study. What's most relevant to us today is to understand that the interest in the construction of a mechanical human is not of modern origin but has its roots in the dawn of time. What is it about this seemingly impossible idea that captures our imaginations, just as it did for our ancient ancestors? Later we will see that this drive has deep cultural origins (see Chapter 2) that still influence the way we perceive and interact with robots, both real and fictional.

It wasn't until 1961 that the robots we know today made their debut on the world stage. During that year, General Motors introduced Unimate, a machine consisting of a mechanical, extendable arm with a gripper, attached to a rotating base.

Unimate's job was to safely grasp die castings from an assembly line and weld them onto car bodies, a dangerous task for human workers that came with significant health risks. Unimate is considered the first industrial robot: a category of machines designed for manufacturing, built and programmed to perform repetitive tasks in highly controlled environments, such as picking an object from a conveyor belt and placing it elsewhere. They often look like mechanical arms of varying levels of complexity where a certain number of fixed parts (known as links) are connected to mobile parts (joints) and to what is called an end effector, which can be a manipulator such as a hand, gripper, or any other tool needed for the robot's task. The combination of these components determines its degrees of freedom: the complexity of the movements it can perform. As a reference, consider the human upper limbs, which consist of three joints (shoulder, elbow, and wrists) connected by two links (arm and forearm) and ending in a manipulator (the hand), providing a total of seven degrees of freedom: one for each direction in which each of our joints can move and rotate. For instance, the hand can move up and down, left and right, and twist both clockwise and anticlockwise.

In the past, industrial machines acted more like automatons rather than intelligent robots. This was because they performed their tasks blindly, meaning that they would follow a predetermined sequence of operations defined by their programming. For instance, they would be instructed to position their hand at specific coordinates, open the gripper, move down, and so on. Nowadays, industrial robots rely on the use of sensors: electronic devices that enable them to analyze their surroundings, allowing them to correct or direct their movements. If an object on the assembly line ends up not being in the expected position, a blind robot would simply grasp the air and continue with its task as if nothing happened, while a modern robot would correctly identify the new position and avoid missing the target.

While industrial robots represent an important technological advancement that has greatly benefited manufacturing and human well-being, this book focuses on a different and more recent kind of robotics that is far less mature due to its increased complexity and scope. Throughout the rest of the book, we will concentrate on robots that are designed to operate in intricate, dynamic environments, often in proximity to humans. These are known as service robots, and they represent an area of robotics that is rapidly evolving and has the potential to revolutionize our daily lives.

There are significant differences between these two categories. Industrial robots are meant to replace humans, operate in well-defined environments, and do not need to make autonomous decisions beyond a few corrections to their trajectories. All necessary information is programmed in advance, and they simply follow instructions. In contrast, service robots are designed to assist humans, must operate in highly unpredictable settings, and require a much higher level of intelligence and adaptation to navigate complex environments.

Imagine a scenario where you have a robot vacuum cleaner that needs to navigate around a house while avoiding obstacles on its path. How can you ensure that the robot doesn't accidentally collide with furniture or fall down a flight of stairs? One might assume that simply providing the robot with a map of the house would suffice, but in reality, it's not that simple. What would happen if a chair was moved into the middle of a room?

What if someone suddenly stepped in front of the robot? Dealing with these unpredictable situations requires constant observation and evaluation of the environment. The robot must take a step forward, assess the state of the world surrounding it, modify its plans if it discovers something unexpected, then repeat the process again and again.

For how advanced this process can seem, it pales in comparison to the level of intelligence required for a machine to engage with humans on a social level. Speech understanding, emotion recognition, and physical collaboration are just a few examples of the complex skills necessary for a robot to navigate our social landscape, one that even humans find challenging at times. In the following chapters, we will explore the field of social robotics, which breaks through the realm of automation and ventures into the realm of psychology.

Having explored the history and narrative arc that have guided our idea of robots through the ages, it's time to wonder: how would we define a robot today? In our modern, science-driven world, one of my favorite definitions comes from Rodney Brooks, roboticist and entrepreneur: a robot is a machine that senses the world around it using sensors, performs some computations, decides a course of action, and enacts it to produce a change in the external world. This last part of the definition is important because it distinguishes robots from devices like dishwashers or washing machines, which can only generate internal changes.

According to this description, a robot is composed of three basic components: sensors, actuators, and a computing unit. As I have already mentioned, sensors are devices that gather information about the environment and convert it into electrical signals. Typical examples are cameras that translate light into digital images and sonars that measure the presence and distance of obstacles. Actuators enable the robot to interact with its surroundings, such as through movement or manipulation. For instance, wheels allow for mobility, and motors control the limbs to perform actions. Finally, the computing unit processes data from the sensors and provides instructions to the actuators, enabling the robot to make decisions and act accordingly. I shall refer to this cycle of sensing, reasoning, and acting as the autonomous robot loop: a pattern that lies at the core of many, if not all, robotic behaviors.

Let's revisit the example of our vacuum cleaner robot, which is programmed to navigate around a home while avoiding obstacles. The robot's first step is to utilize its sensors, such as a sonar. This device, inspired by the sensory systems of bats and dolphins, emits an inaudible sound wave that travels forward and bounces off nearby surfaces. If the sensor detects a returning signal, it assumes the presence of an obstacle. To determine the distance of the latter, the robot measures the time it takes for the sound wave to return. The longer the time, the farther away the obstacle is.

Next, the robot's onboard computer processes this information and decides whether the distance of the obstacle is within an acceptable range. For example, it may assume that an obstacle is too close if it's no more than 20 cm away. In that case, the computer will send a command to the motors that control the robot's wheels, causing it to steer away from the obstacle.

After this action is taken, the autonomous robot loop begins anew, with the robot continually sensing its environment and adjusting accordingly. By adapting to changes in its surroundings, the robot is able to navigate effectively while avoiding collisions with obstacles.

Now that we have established what constitutes a robot and explored its various applications, let's delve into one of its most notable features: its appearance. Are a robot's looks purely aesthetic, or do they serve a functional purpose as part of its design?

In our image

Even though robots come in a wide variety of forms, a quick Google Image search for the term "robot" reveals that our expectations about these machines are often shaped by depictions in fiction, where they are typically portrayed as human-like machines. This trend can be traced back to science fiction classics such as *Terminator*, *I Robot*, or *Star Trek*. Among these, *R.U.R.* was especially significant as it introduced us to the struggle of dealing with emerging technology that looks and acts like humans but ultimately is not human.

Robots depicted in literature and entertainment offer us a glimpse into our own nature. They allow us to reflect with detachment on topics such as intelligence, humanity, and slavery. In these stories, a human-like body represents the idea of personhood. Consider the differences between Data from *Star Trek: The Next Generation* and R2-D2 from *Star Wars*: the former has a human body and struggles to become one, while the latter does not. This difference is not coincidental but reflects their intended roles and purposes.

The driving idea is that we tend to associate human-level intelligence with human-shaped bodies, in other words, it is easy to assume that you need a human to do a human's job. In the real world, there is some truth in this, but also some caveats. It is true, for example, that we live in a human-shaped world, where everything is designed to be used by a human-like body. Think about cars: if a quadruped alien would ever set foot on Earth, it would have a very hard time trying to operate one of them! This means that having an anthropomorphic (human-like) form would help a robot to blend in our environments, where it is expected to operate. However, if a robot is designed for a specific task, it may be better to engineer its form to optimize performance. For example, search and rescue robots meant to navigate through rubble and tough terrain are usually designed with sturdy crawler wheels rather than legs. The same reasoning applies to the design of space robots, such as the Mars rovers that face the harshness of the Red Planet's surface.

A robot designed to resemble a human is called an android or gynoid, depending on whether it aims to mimic a masculine or feminine appearance. These machines strive to imitate human beings as closely as possible and use artificial skin, synthetic hair, and other means to do so. Since they are intended to appear as realistic as possible, they often feature artificial facial musculature to simulate emotional expressions and other movements related to speech. Androids and gynoids are the type of robots that science fiction writers often use to explore the meaning of humanity. Characters such as Deckard in *Blade Runner* and Andrew in *Bicentennial Man* serve the purpose of investigating curiosity, emotions, and the line between human and non-human. From this perspective, the inherent duality of these characters turns into a dramatic tension between their two natures that meet and clash during a journey of discovery, both for themselves and for the audience.

From a real-world perspective, robots designed to resemble humans serve two main purposes. The first is entertainment, as seen with Erica, a gynoid developed by Hiroshi Ishiguro at Osaka

University, Japan, for the movie *B*. More commonly, though, these robots are used for social purposes, such as conversing with people. Sophia, a gynoid created by Hanson Robotics, is a notable example that garnered significant media attention when Saudi Arabia granted it citizenship in 2017 (Figure 1.2). However, this decision was met with criticism from the robotics community, who accused the creators of exaggerating and misleading the public about the robot's true capabilities.

Figure 1.2 Shopia, a gynoid robot developed by Hanson Robotics. International Telecommunication Union, CC BY 2.0, via Wikimedia Commons.

A machine that is designed to have a human-like shape without necessarily attempting to mimic a human is known as a humanoid. These robots typically have a head, torso, and two pairs of limbs, although not all parts of the body need to be present. For example, Pepper, a well-known social robot developed by Aldebaran, has a humanoid upper body that fits onto a mobile, wheeled base resembling the tail of a mermaid (Figure 1.3). Generally, any robot with an anthropomorphic upper body can be classified as humanoid, regardless of its appearance. For instance, Pepper's head has an elliptical shape with two large, sunken black eyes, and no nose. While it is not human-like in appearance, its shape is unmistakably humanoid.

As previously mentioned, the primary reason for designing a robot with a humanoid body is to provide it with an advantage in handling tools and operating in environments meant for humans, such as homes or offices. However, there is another reason to consider. A robot that resembles a person or a living being can trick our minds into attributing it with qualities we typically associate with biological organisms. This can have a profound impact on our perception of these robots and how we interact with them, including our expectations of their functionality. Consequently, this can affect other factors such as our decision to trust or distrust these intelligent machines. We will delve into this topic further in Humans trusting robots (Chapter 2).

There is another significant reason that drives us to study and develop humanoid robots, which we have already touched upon. The field was originally born with the purpose of engineering better prosthetic limbs, such as hands and legs, the design of which requires a comprehensive understanding of the human body. As knowledge in this area has progressed over time, it has led to advancements in technology, such as the development of active devices that not only support the body but also contribute to its function (e.g., powered legs or feet that aid users in climbing stairs). A close collaboration between biomedical and

Figure 1.3 Pepper, a humanoid social robot developed by Aldebaran.

engineering disciplines quickly emerged, where advancements in one field would lead to progress in the other and vice versa. As the creation of complete artificial bodies became possible, this relationship was further strengthened and extended to include social sciences such as psychology and cognitive science. This interconnection allows scientists of these disciplines to test their hypothesis on robots and achieve an improved understanding of human beings, which leads to the design of better robots. Once again, we are witnessing a virtuous circle between robotics and other domains of science.

Some people believe that the connection between robotics and human knowledge can be taken to the extreme by using advanced robotics to enhance human capabilities. This is the fundamental idea behind transhumanism, a philosophical movement that asserts humans can and should evolve beyond their current physical and mental limitations through scientific and technological innovation. While this is an interesting and complex topic that is best explored in a dedicated book, it is worth mentioning for the benefit of robotics and technology enthusiasts.

Now that we have a general understanding of the external appearance of robots, it's time to take a closer look. Let's peek inside of them to probe the workings of their electronic brains and see how they differ from the human ones.

What's in your mind?

Although the concept of intelligent mechanical servants has been around for several millennia, the term "artificial intelligence" (AI) was not coined until 1956. It was during this year that American computer scientist John McCarthy assembled a group of researchers from various disciplines in Dartmouth,

U.S., to discuss and exchange ideas about thinking machines. This umbrella term encompasses a set of skills that we normally associate with human intelligence: the ability to understand, to reason, to make decisions, and to learn from experiences. However, can we truly replicate the complexities of the human brain in a machine?

This question was famously raised by British mathematician Alan Turing in the mid-20th century in his famous paper *Computing Machinery and Intelligence* [5] which opens with the following inquiry: can machines think? In this manuscript, Turing introduces a game originally known as the Imitation Game, now commonly referred to as the Turing Test. The game involves a human subject (the "interrogator") in communication with other two entities: one of them is another human, while the other is a computer. Since there is no direct line of sight between the interrogator and the other two entities, all communication occurs through text. The purpose of the game is for the interrogator to distinguish the human from the machine, despite the best efforts of the computer to disguise itself as a person. According to Turing, a computer capable of successfully impersonating a human can be considered capable of thinking like one. This is because it would have to display inherently human capabilities, first and foremost the capacity to hold a conversation, including both language comprehension and synthesis.

Of course, this experiment is not without its flaws, the most notable of which is exposed by philosopher John Searle's Chinese Room argument, formulated in 1980. This thought experiment aims to demonstrate that a computer subjected to the Turing Test does not truly understand the conversation, but rather only responds mechanically to its programming. Searle imagines himself locked in a room full of Chinese characters he cannot understand, along with a book of instructions written in English, his native language. If a Chinese speaker outside the room sends him a message, he can use the manual to select an appropriate response. Even though Searle cannot understand a single word of the Chinese language, the person outside the room would believe that they are conversing with someone who does. In other words, the person in the room could pass a Turing test as a Chinese speaker without truly understanding any of the conversation!

In the summer of 2022, former Google engineer Blake Lemoine made a bold claim that their AI-driven chatbot, LaMDA, had achieved sentience. Lemoine based this assertion on a series of conversations he had with the machine, which led him to believe that it had become self-aware. The story quickly gained traction in the media (in part due to sensationalistic journalism), and within days, everybody was talking about it. The scientific community largely rejected Lemoine's claims, and the main reason boiled down to the Chinese Room argument: a program developed to manipulate linguistic symbols does not truly understand the language it is using.

Despite all of this, Turing never intended for his test to be a practical benchmark for evaluating machine intelligence. Rather, he developed it more as an easily understandable example that would contribute to the ongoing debate about the philosophy of AI. In fact, researchers do not use it as an evaluation tool for their experiments. The reason behind this is explained by Stuart Russell and Peter Norvig, authors of a widely influential entry-level textbook on AI [6], with a clever analogy: airplanes are tested for their flying capabilities, not for how closely they can resemble birds. However, this doesn't mean that the Turing Test was never of any practical use. For example, it lived until 2020 in the form of the Loebner Prize: an annual contest that assigned a prize to the most human-like chatbots (computer programs designed to simulate conversations with human users). American non-fiction author Brian Christian has written an interesting book on the matter, reflecting on how our idea of humanity is challenged by the rise and evolution of computer systems that imitate us [7]. He argues that AI programs are taking a foothold on shores that were previously considered exclusive of human beings, such as reasoning and conversation, forcing us to rethink what makes us humans. For example, what does the existence of ChatGPT tell us about our personhood? Does creating a machine capable of emulating language, one of the hallmarks of humanity, make us less unique in the grand scheme of the universe?

Thanks to his Chinese Room argument, Searle was the first to draw a distinction between weak AI, a computer program able to simulate only a narrow range of human mind functions, and strong AI, characterized by full human-like mental capabilities. As of today, the latter remains a hypothetical concept, and the AI systems that exist are specialized and capable of executing only specific tasks. For instance, it wouldn't be possible to expect a chess-playing program to be able to handle a game of checkers. Nonetheless, even weak AI has had a deep and pervasive impact on our society, revolutionizing many aspects of our lives, some of which we may not even be aware of.

Let's set an example to paint the bigger picture. You take out your brand-new smartphone and try to unlock it. The camera utilizes an intelligent face detection program to determine if you're the owner of the device. Once you gain access, you open your go-to social media app and, again, it is an AI working in the background that decides which content (and, mostly, ads) you see. You then upload a picture of your delicious meal and select a filter for it, which is applied to the image by a dedicated AI. Finally, while adding a caption, you are assisted by an intelligent auto-correct and word suggestion software. Hopefully, this should give you an idea of the extent to which AI has become integrated into our society and the many ways in which it impacts our daily experiences.

While we are still far from the sentient machines depicted in science fiction, the impact of weak AI is tremendous. In just a short span of time, it has changed the way in which we think about and solve problems, and has opened the gates to new, exciting possibilities. One could just imagine the consequences that a strong AI could have on our world! In 2010, *Scientific American* published a list of twelve possible future events that could drastically alter the world [8]. This list comprised both natural and human-made scenarios, spanning from asteroid collisions to nuclear warfare, melting ice caps, and even the discovery of extraterrestrial life. Notably, one item on the list was the hypothetical uprising of self-aware machines, claiming their rightful place in the world. Only a few years later, in 2015, a group of scientists including Stephen Hawking signed an open letter in which they warned the scientific community about the dangers of unregulated AI. Since then, the debate on AI ethics has continued among scientists, policymakers, stakeholders, and the general public. While a robot uprising may not be imminent, it is essential to maintain the conversation around AI ethics to ensure that future developments are used for the greater good of humanity.

Why has it not been possible, as of today, to develop a strong AI? There is a paradox known in the research community, called Moravec's Paradox, which states that research in AI has been able to create computers that can perform tasks that are difficult for humans, such as reasoning on complex mathematical operations, but that struggle with tasks that we consider easy, such as folding the laundry. We tend to take for granted the power of our cognitive functions and often forget that behind mental feats that take place in the span of milliseconds, there is a complex and intricate organ that we still don't fully understand. One significant obstacle is the definition of intelligence itself. Early assumptions were that a machine could be considered intelligent if it displayed cognitive skills associated with the human mind. However, science has since moved forward, and it is now recognized that merely mimicking human abilities is not enough. A strong AI should be able to behave rationally, acting as a truly autonomous agent that defines its own goals (something known as intrinsic motivation) and tries to achieve them. However, as of today such an entity remains purely theoretical.

At this point, you might be wondering how an AI actually works under the hood. There are two main schools of thought on this topic, which we shall refer to as the symbolic and sub-symbolic approaches.

Symbolic AIs were the first to be developed, back in the mid-1950s. To understand how these programs work, imagine an abstract land made up of small basic blocks, which we call symbols. Each symbol represents a concept in the real world: either physical entities (cats, people, houses), abstract concepts (web pages, money transactions, a degree), actions (cooking, teaching), or even states (being taller than something else).

Symbols are involved in various types of relationships with each other and can be manipulated to create more complex structures. For example, to express the concept of "the cat is on the table" we would need three symbols: one for the cat, one for the table, and another one to represent the concept of "being on top of something." Then, we could represent that statement as such:

$$On\ (Cat,\ Table)$$

This notation tells us that the entity "cat" is in a relationship "on top" with another entity named "table." Statements like this enable a symbolic AI to formalize the current state of the world and determine which actions are necessary to transform it into the desired one, much like a puzzle solver trying to figure out the best moves to reach the end goal. By breaking down the world into small blocks of symbols, the AI can reason about complex concepts and relationships in a way that simulates human thought.

For example, a symbolic AI program to keep the cat off the table could look something like this:

If the current state is:	$On\ (Cat,\ Table)$
Then perform the action:	$Place\ (Cat,\ Floor)$
To obtain the resulting state:	$On\ (Cat,\ Floor)$

A process that acts within this abstract world must have the ability to create, reshape, and destroy these symbolic structures in order to reason and solve problems: that is the essence of a symbolic agent.

Sub-symbolic AIs, on the contrary, take a completely different approach, disregarding symbols in favor of a more mathematical and implicit representation of reality. The jewel in the crown of these techniques is the artificial neural network, a computer system that is inspired by biology. In natural systems, the brain is made up of neurons, special cells connected in a network that conducts electrical messages throughout the organ. Each neuron is part of a chain: it receives impulses from other cells and, based on the strength of the stimulation, decides whether to send its own impulse along the network (this is called "firing"). The trick here is that the connections between them are not all the same, rather some links can be stronger or weaker than others: this means that a neuron could be more easily stimulated by one or more neurons than others. In order to learn something, the brain changes itself by altering the strength of these connections, in fact rewiring itself.

Artificial neural networks work on the same principle and are used in tasks such as pattern recognition and classification: the act of identifying objects belonging to different categories. To train the network to distinguish, for example, cats from dogs, it is shown one picture at a time and its response is gradually corrected until it rewires itself enough to effectively separate the two categories. This is the most famous (but not the only) paradigm in artificial neural networks, known as supervised learning. This name derives from the fact that the network is provided with the correct answer for each training example it is exposed to, like a student learning from a teacher's demonstration.

Up to this point, we have examined robotics and AI as two separate entities, each with its own unique history and development. Eventually, as we expect, these two disciplines reach a point of contact, and their marriage produces some of the most exciting technological marvels of this century. What is then the relationship between robotics and AI? Can one exist without the other?

The ghost in the machine

As we have seen, it is possible for AI to be disembodied, which means to exist without a physical form as a pure computer program: a chess-playing software doesn't require a body, rather, it can carry out its tasks in a pure virtual environment. Garry Kasparov, the world-renowned chess grandmaster, experienced this firsthand when he was defeated by the Deep Blue supercomputer in 1997. On the other hand, a robot does not have to be intelligent to perform its duties. Industrial robotic arms, for instance, are programmed to perform highly repetitive and controlled movements on a factory line. Thus, while AI and robotics are often thought of as interrelated, they can also exist independently of each other.

The moment in which we bridge these two disciplines is when things start to get interesting. By doing so, we step into the domain of artificially intelligent robots. We previously discussed how industrial robots reached a turning point when they began to be equipped with sensors that allowed them to perform basic reasoning tasks. A camera, for instance, can enable them to detect and correct displacement errors on the parts they are working on. This demonstrates how adding even a small amount of intelligence to these robots allows them to perform impressive actions, such as saving lives by avoiding collisions with people who might carelessly walk too close to them.

Intelligence, both natural and artificial, is not a binary quality that is either present or missing: rather, it manifests as a spectrum that allows varying degrees of mental capabilities. But how can we measure intelligence when it's so difficult to define in the first place? Researchers and engineers often associate greater mental capabilities in machines with a reduced need for human intervention during their operation, and vice versa. To better understand this, let's look at the two ends of the spectrum. On the lower end, we have telecontrolled (or teleoperated) devices that are manipulated step-by-step by an operator. Notable examples of these are surgical robots, which are remotely controlled by a surgeon to perform medical operations. The most famous surgical robot is Da Vinci, a marvel of modern technology developed by the American corporation Intuitive Surgical and used worldwide. On the opposite end of the spectrum, we have super-intelligent, fully autonomous, and independent robots, which, as of today, only exist in science fiction.

Between these two extremes lies the category of semi-autonomous robots, which includes a wide range of machines with different capabilities. Some of the most renowned machines in this category are known as reactive robots. These machines can sense their surroundings and react accordingly, but they lack any memory of the past and only respond to present stimuli. In his book *Vehicles: Experiments in Synthetic Psychology* [9], Italian neuroscientist and cyberneticist Valentino Braitenberg described some thought experiments involving these machines, which are now known as Braitenberg vehicles (Figure 1.4). His goal was to illustrate how complex behaviors, such as basic forms of cognition, can emerge from simple systems. His imaginary robots cannot reason, instead they operate purely on instinct: their light sensors are directly connected to their wheels, causing them to move faster in response to stronger stimuli. Braitenberg demonstrated that this simple structure can simulate some simple emotions, such as "aggression" (the robot runs towards a light source) or "fear" (the robot tries to escape into the darkness).

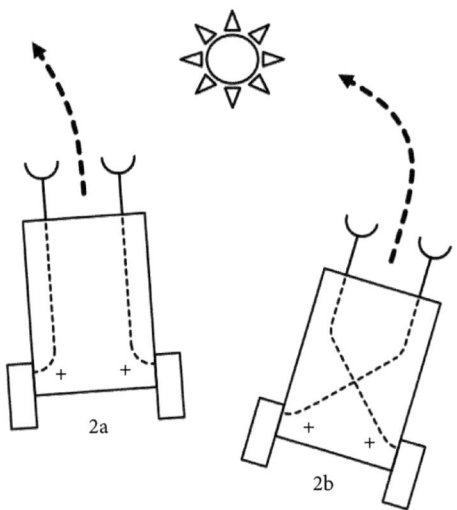

Figure 1.4 Example of Braintenberg's vehicles. The robot named "2a" enacts fear, as the wiring of its light sensor make it steer away from the light. In contrast, robot "2b" is attracted towards the light and will crush into it, therefore enacting "aggression." Thomas Schoch – www.retas.de, CC BY-SA 3.0, via Wikimedia Commons.

It should be clear by now that researchers develop artificially intelligent robots to allow them to perform tasks and adopt behaviors that are beyond the capabilities of pure mechanical control. However, there is another critical reason to do so, which is particularly relevant to the purpose of this book. To fully understand this reason, we need to delve into the realm of philosophy and discuss a topic that has engaged great thinkers since the beginning of time: the mind–body problem.

The latter is an age-old debate concerning the relationship between the mental and physical aspects of human experience. In other words, are mind and brain two separate entities, and if so, what is their nature? This debate is simply formulated but has proven to be incredibly difficult to solve. The fundamental question is: how does the complexity of the human mind arise from a collection of neurons and gray matter? Do thoughts and consciousness exist independently of the physical substance that contains them?

The earliest idea, which is now outdated, was that there was no dichotomy between mind and body, rather, everything was a unified whole. For example, the Greek philosopher Aristotle believed that everything was physical, and mental states were simply different material configurations of the brain. He compared the latter to a lump of clay that could be shaped into different forms. During the 17th century, French mathematician and philosopher René Descartes came up with an idea that is known as Cartesian dualism. According to Descartes, the world was composed of two distinct and incompatible kinds of substance: *res extensa* and *res cogitans*. The former represents the physical realm, which is limited and unconscious, while the latter encompasses the mental realm, which is free and conscious. Descartes believed that the mind could occasionally influence the body, for example, by making us walk towards the fridge when we feel hungry, but not vice versa. British philosopher Gilbert Ryle would later criticize this stance, as it considered the mind as a "ghost in the machine": a non-physical entity that inhabits and controls the brain.

Today, we have a very different understanding about the mind–body problem. Consensus in the fields of psychology and cognitive science highlights the fact that both mind and body can influence each other, and their interaction is so intimate that it shapes their development. The brain acquires experiences from the world and learns from them, thereby expanding its mental capabilities. However, these experiences are only possible because the brain is embedded in a physical form that has the capacity to perceive and interact with the environment. This implies that a body with different capabilities, like that of a bat, would gather unique experiences, shaping its mind in a distinctly different way.

To fully appreciate the power of this concept, let's consider the case of Neil Harbisson, a color-blind British artist who, in 2004, decided to implant an antenna on his skull. This technological implant converts colors into sounds and transmits them directly to his brain. In his captivating TED talk [10], Harbisson recounts his story: he originally viewed the world in black and white and had no concept of color until his device allowed him to perceive it as sound. Initially, he had to undergo a learning phase where he matched each color (which he couldn't distinguish with his eyes) to the sound it produced. But after a brief period, his brain adapted to this new perception and made sense of it. Suddenly, he could not only distinguish colors based on the auditory feedback they gave him, but he also started to dream in color, proving that his brain had fully embraced his new capabilities and was working in synergy with the implant. This was a turning point in his life. Harbisson's fashion sense changed too: he now chooses clothing based on how well they sound together, a concept that feels alien to those of us who don't share his experience. He can hear paintings, and visiting an art gallery feels like walking through a musical concert. Additionally, he developed the ability to find similarities between people's faces by comparing the way they sound to him.

Neil Harbisson is a living example of the power of the relationship between the body and mind and how they are not two distinct entities.

If our goal is to develop human-like AI, we must embed it in a human-like body that allows it to share the same experiences that have driven us to become what we are. This concept is summarized by the notions of embodied intelligence (the idea that an intelligent agent is not just a disembodied mind but has a physical body through which it interacts with the environment) and situatedness (the idea that an intelligent agent's behavior and cognitive processes are heavily influenced by the specific environment or context in which it is located).

While all of this sounds great in theory, it's important to understand what it means in practice. Let's examine a real-world example of how a robot can use its body to develop capabilities, inspired by how human infants learn how to move around. Newborns have no idea how to use their bodies to their advantage. At this stage of life, babies perform seemingly random movements with their arms and legs, which are jerky and reflex-like. As they age, these movements become more precise and directed, until the baby becomes capable of reaching for objects and crawling. With time, infants learn to use their bodies effectively by practicing and becoming more familiar with them. A robot can use the same approach to learn how to perform movements, using an exploratory strategy in which the machine gradually learns the effect of moving its limbs. The robot can then use this acquired knowledge to move in space and achieve its goals. Of course, this process is made possible by the close interaction between the mind (the AI program) and the robotic body.

A study from the University of Plymouth, UK [11] showed how a little humanoid robot was able to use this technique (named "motor babbling") to autonomously learn how to tilt on each of its sides. The kind of robotics that draws inspiration from the developmental process of infants and children as they mature into adulthood is known as developmental robotics.

Of course, it would have been possible to write a program containing all the information necessary to perform the desired movements and call it a day. However, what would we have learned from that, apart from some nifty mathematics that proved to work? Remember about the virtuous circle of robotics: we study robots to better understand humans and, in turn, develop better robots. If this is our main goal, we must strive to make robots learn in the same way humans do, to bridge the gap between the simple artificial minds we are able to create today and the kaleidoscopic complexity of the human being.

Early computers developed in the 1940s were the size of entire rooms. If you had told someone back then that in less than a century everyone would always carry multiple computers on them, they would have laughed at you. Yet, here we are, surrounded by them: there's probably at least one in your pocket, one on your wrist, and one on your desk. You might even be reading this book on a screen right now.

Just as computers have evolved from science fiction dreams to everyday reality, robots are following a similar trajectory. Once confined to factories and the pages of sci-fi novels, robots are becoming increasingly integrated into our lives. They are already making appearances in public spaces and many people have had at least one interaction with them. By following this parallel, I predict that in just a few decades we will be living amongst robots: they will be delivering the mail, cleaning the house, offering companionship, and even acting as teaching support in schools. To achieve this level of integration in our society, we cannot expect people to learn how to communicate with robots. In other words, we don't want to robotize people, rather, we hope to make the minds of these mechanical companions a little more human.

This concludes our quick introduction to the fascinating world of robotics. As you've seen, robotics is a multidisciplinary field that encompasses computer science, mechanics, electronics, history, anthropology, psychology, and cognitive science, among others. Each of these subjects was covered very briefly and deserves a book of its own to be fully appreciated. This one, though, has a specific focus: exploring the topic of trust between humans and robots, a fascinating and complex matter in its own right. Let's begin this exploration, a journey of discovery into the human mind, the nature of AI, and the ways in which these two realms intersect. Through this exploration, we hope to better understand the relationship between humans and robots, and how we can work together to build a more advanced and interconnected society for the future.

CHAPTER 2

HUMANS TRUSTING ROBOTS

Matter over mind

This book focuses on the scientific discipline of human–robot interaction (HRI), a branch of robotics that studies, maybe unsurprisingly, the interactions that happen between humans and robots. It is a multidisciplinary field that not only considers the technical aspects of robotics and AI but also welcomes contributions from psychology, social sciences, and design. Its main objective is to provide guidelines for creating robots that possess the necessary skills and characteristics to enhance their interaction capabilities with humans.

This is a relatively new field that has emerged since the mid-1990s and early 2000s, rising on the foundations laid by its predecessors. The oldest of these is a scientific subject known as anthroposemiotics or, more commonly, human communication. This discipline is concerned with understanding the complex and multifaceted nature of human verbal and non-verbal interaction.

Many areas of science have rich and layered backgrounds that we might take for granted, often without realizing it. At first glance, communication appears straightforward: one can approach another person, gain their attention, and begin conveying ideas and information using a shared language. However, there is a great deal more happening beneath the surface that we are often unaware of. Effective communication requires us to take the other person's perspective into account. We must consider their level of understanding, determine how much background information to share, and be mindful of any cultural boundaries that may exist. However, this information can only be guessed, as no one can directly access another person's thoughts and feelings. In order to enable effective interaction, it is first and foremost necessary to establish a common ground that enables the sharing of mindsets and viewpoints. Communication can then be defined as a shared and symbolic interaction (we have already discussed the meaning of a symbol in Chapter 1, in the context of symbolic AI). Anthroposemiotics explores the various expressions of human communication, including rhetoric, and aims to understand the ways in which we transmit information to each other.

This discipline informs us about what occurs in interactions between humans, but in the 21st century communication is no longer something that involves exclusively people or, for that matter, purely living entities. On the contrary, many of us spend more time communicating with computers than with other people, especially during the long lockdowns that we have recently endured worldwide. The question then becomes: how can we effectively communicate with these machines? Humans and computers are fundamentally different entities that speak distinct and often incompatible languages, and even reason in vastly dissimilar ways. Until the middle of the last century, we used to communicate with computers using punch cards: pieces of paper that could be punched in predefined positions to represent binary information, which the machine could read and interpret as instructions. Creating these cards was difficult and tedious. Engineers used to spend hours manually punching holes into cards, with no way to detect errors until the entire card was completed and inserted into the computer. If a mistake was found, they often had to redo the entire card, with significant time loss and frustration. This form of communication was imbalanced, heavily favoring the machine and straying far from the norms of regular human exchanges. Could there be a way to make computers more human-like so that they could interact with us naturally, rather than forcing us to communicate like them? Could we create a shared workspace with computers, allowing us to exchange information as seamlessly as we do with other people?

These questions led to the foundation of the field of human–computer interaction (HCI), which was formally established in the 1983 book *The Psychology of Human-Computer Interaction* edited by Stuart K. Card [12]. The main purpose of this field is to study optimal interfaces between people and computers. An interface, in this context, is any possible point of contact between the human and the machine. For example, input and output devices such as keyboards, mice, and monitors, or graphical software interfaces that allow us to interact with the computer. In HCI, the term "computer" encompasses a wide range of devices that possess computing power, such as laptops, smartphones, gaming consoles, televisions, and more.

HCI experts bring together knowledge from both computer and social sciences to design systems that are intuitive to use. These professionals must possess skills in domains such as computer graphics, design, cognitive psychology, and linguistics. Their goal is to ensure a good level of usability, which is a measure of how well an operator can use a product, whether it's a webpage or a voice assistant, to achieve their objectives. Humans are lazy and emotional creatures that are prone to frustration when something doesn't work as they expect it to. How many times have you found yourself searching every corner of a website to find the button that would allow you to cancel a subscription, almost giving up in frustration? It turns out that this is often done intentionally: it's a specific design pattern that exploits our inherent laziness in an attempt to retain our membership and money. On the diametrically opposite end of the spectrum, it's easy to recognize when an interface has been intelligently designed: everything is clear and readable, you know exactly what you should be doing, and the interaction feels natural and pleasant.

To showcase how bad design can negatively impact the user experience, in 2017, a group of designers gathered on Reddit, one of the most popular social forums on the internet, to propose the worst possible volume control interface ideas they could come up with. The results of this experiment were both hilarious and informative, with ideas ranging from slingshot controls that required the user to charge and release to hit the desired volume level, to random controls based on rolling a set of dice [13].

While HCI's primary goal is to make our lives more comfortable, its impact goes far beyond convenience. In fact, our safety can often depend on how efficiently we can interact with the technology around us. A striking example of this occurred in Pennsylvania, U.S.A., in 1979. In the early morning of March 28, in the brisk hours that precede sunrise, no one would have suspected that a disaster was brewing at the Three Mile Island Nuclear Generating Station. At around 4 a.m., a malfunction in the cooling system caused a partial meltdown of the Unit 2 reactor core, which in turn resulted in the release of large quantities of radioactive material in the surrounding area. Subsequent investigations determined that the incident was at least partially caused by the inadequate human–machine interaction design of the control room and computer interfaces. Apparently, the panels were arranged inconveniently, the alarms were ambiguous and too similar to one another, and the instruments on the console were not reporting accurate measurements about the cooling unit. All of this caused the operators to fail to recognize the ongoing emergency, leading to the catastrophe and the subsequent decommissioning of the station.

The invention of HCI was driven by the growing presence of computing machines in our lives. Similarly, at the end of the last century, scientific progress in the fields of robotics and artificial intelligence led to the emergence of HRI, a thriving discipline firmly grounded in its predecessor. But why did we need to create a new subject in the first place? Can't robots be treated like any other computing device? Isn't a robot essentially a computer on wheels?

If you have been following along with Chapter 1, you already know that a robot is much more than a mobile computer that moves around the house. Instead, it is an embodied agent: an entity that can interact with its environment through a physical body. Because robots have a bodily presence around us, they provoke a whole constellation of psychological, emotional, and moral responses that a normal appliance wouldn't elicit.

To help you understand the full extent of what we are talking about, let me tell you a story that took place in 2020. Several Italian news sites reported that an engineer working in the town of Pesaro became enraged while at work and began destroying everything in his path, including computers and printers. By the time the local police forces managed to calm him down, the office was left in a state of chaos, with broken equipment scattered everywhere. Leaving aside any considerations about the employee's temper, I would like you to take a moment to reflect on and answer the following question: what are your feelings towards the mistreated computers lying on the floor in pieces?

The question might have startled you. It's not easy to give a straightforward answer, because we don't usually attach any emotion to objects, especially when they don't belong to us. You might experience a sense of pity towards the materialistic and senseless loss of expensive equipment, but probably not much else.

Now let's consider another case. You may be familiar with Boston Dynamics, a famous robotics company based in the United States that often releases viral videos on social media platforms. In one of these [14] Spot, one of their robot dogs, is seen standing quietly in a parking lot when suddenly a man approaches and kicks it on its side. The robot staggers, sliding on the road covered in a thin layer of frost, struggling to keep its balance. It moves its legs frantically, trying to prevent the fall, and after a few moments of uncertainty, it finally manages to regain its posture. Take a moment to picture this scene in your mind or, even better, watch the video and consider the following question once again: how do you feel towards the mistreated robot?

If you have experienced a sense of inexplicable outrage towards the poor robot's struggle, it might comfort you to know that you are not alone: many among the 31 million viewers who watched the video shared that same feeling. The short film caused such a reaction when it was released in 2015, that PETA, one of the largest animal rights organizations in the world, had to remind people that, despite how inappropriate kicking a robot dog might feel, Spot is still a machine with no consciousness and no sense of pain.

There is clearly something going on here. Why is it ok to break a computer but not to kick a robot? Something must be happening in the background of our minds, influencing our reactions.

That something has a name: anthropomorphization, a word that derives from the Greek *ánthrōpos* ("human") and *morphē* ("form") and describes the act of endowing something with human qualities. We are incredibly good at doing this: we constantly project human attributes and traits to entities belonging to both the natural and the artificial world, such as animals and robots. Consider our stories and myths, where elements of the natural world embody aspects of humanity: owls are wise, foxes are cunning, and snakes are malicious. Raise your hand if you have never smiled at the sight of a dog dressed in clothes, be it a formal suit or a superhero cape. This cognitive process is intrinsic to human nature and involves projecting human needs and explanations onto the behavior of non-human entities. It's a mechanism that helps us make sense of the world and creates a deeper connection with our surroundings. However, it can also bias our understanding of other entities, like Spot the robot dog. Despite our rational knowledge that Spot is a machine with no consciousness, merely following its programming, we still experience empathy when it displays traits associated with life.

According to evolutionary psychologists, anthropomorphization is a byproduct of evolution. Our brains have evolved to process social information so quickly that we tend to over-attribute traits and detect meaningful patterns even in random noise, such as when we see shapes in the clouds.

Whatever the case may be, robot makers are aware of this predisposition and utilize it in the design of their machines. It might sound mischievous, but it is actually used for a noble purpose: by making robots act and look in a way that facilitates the projection of human characteristics, they ensure a more natural and pleasant user experience. In essence, our tendency to anthropomorphize is utilized to increase the acceptability of robots and our disposition towards them.

The example provided has hopefully persuaded you about the substantial differences between HCI and HRI and why the latter requires a distinct field of study due to the unique qualities and capabilities of robots. Scientific breakthroughs in this discipline are applied to a wide range of practical applications: for instance, to develop companion robots that can take care of elderly people suffering with loneliness and cognitive impairment, to help autistic children forge social connections with their peers, to foster human–robot collaboration both in industrial settings and during activities such as driving (autonomous cars are considered robots) and even to support space exploration. In whichever situation we might think of teaming together humans and robots, HRI informs us about how to design human-friendly and useful artificial partners.

The essence of trust

Today's society is the result of millennia of human civilization, and it is a complex and stratified one. We have become accustomed to relying on others to solve many of our problems, such as not having to hunt or gather ourselves to put food on our table. However, in doing so, we often forget the profound interdependence we have with others, whether they are people we know, strangers, or institutions. To bring a meal to your plate, a whole crowd of people must work in synergy. Supermarket employees, producers, truck drivers, health inspectors, and regulators are just a few examples. Each of these individuals operates independently of your control, making it impossible for you to personally verify that the producer did not use toxic chemicals on the produce, that the health inspector was not influenced by the food company, or that the policies established by the regulators are the best ones for your health. Instead, all you can do is place your trust in the competence and dedication of each of these individuals. In our modern society, it is simply not feasible to do everything by ourselves or to verify all the evidence we are asked to believe. To navigate through daily life, we are required to trust others, but also to know when not to trust. Without these capabilities, our lives would become significantly more challenging, if not impossible.

Since trust turns out to be so essential to our lives, let's try to define it. We soon discover that our attempt is hindered by the same problem we have encountered when we tried to define intelligence in Chapter 1: it is a broad notion that spans different aspects of reality, making it difficult to frame. Just think about the different meanings that trust has in the context of personal relationships, economy, or cybersecurity.

Despite these challenges, Mayer, Davis, and Schoorman were able to develop a compelling definition in their 1995 paper *An Integrative Model of Organizational Trust* [15]: "trust is the willingness to be vulnerable to the actions of another party based on the expectation that the other will perform a particular action important to the trustor, irrespective of the ability to monitor or control that other party."

Let's break this down. To establish a trust relationship, there must be two parties: the trustor who grants it and the trustee who receives it. While trust can be a bidirectional relationship, we will focus on the simpler, unidirectional case for now. The definition of trust suggests that it is not a generic relationship between two parties, but rather it is specifically related to some action that is important to the trustor. For example, a parent may trust their child to do their homework on their own but not trust them to go to bed early voluntarily. Moreover, the trustor is exposing a vulnerability by relinquishing control over an important matter and accepting the potential risks and consequences that may result from the trustee not acting in their best interests.

When we take a bus, we inherently trust the driver's skill and goodwill to take us safely to our destination. When we shop online, we trust the company to deliver the advertised product. We trust our best friends to keep our secrets.

Let's consider another example: imagine traveling to the airport to catch a flight and deciding to rely solely on digital copies of your required documents, such as your boarding pass. By deciding to carry no printouts, you are inherently expecting your device not to fail or lose your data. The consequences of this happening would be dire. For all intents and purposes, you are acting as a trustor towards your phone. However, can inanimate objects truly be the subjects of someone's trust?

At a very basic level, we do trust or distrust the objects around us, but in a much different way than we would trust or distrust a person. If your smartphone crashes and refuses to turn on while you are waiting in line at the airport check-in desk, you might feel angry and frustrated, but discovering that your best friend has revealed your secrets to others would be a much more complex and painful emotion, evoking a sense of betrayal. The fact that we would not feel betrayed, but maybe just annoyed, by a smartphone refusing to turn on or by a broken doorbell, indicates that trust in inanimate objects manifests differently from trust in people: to trust an object is simply to rely on it.

Reliability is a characteristic of someone or something that consistently displays the same type of behavior. For instance, an alarm clock is reliable if it never fails to ring at the desired time, a shelf is reliable if it can support a certain amount of weight without collapsing, and a colleague is reliable if they always share fruit with others during lunch breaks. Being reliable is not always a positive quality, as it only indicates a certain pattern of regularity. For instance, consider a friend who is always late, no matter the situation. This person is reliable in the sense that it is possible to predict with reasonable confidence that they will not be punctual at your next appointment, but this behavior is probably far from desirable.

While reliability is all we need when it comes to inanimate objects, it may not be sufficient to trust another human being.

To explore this concept, let's consider two different scenarios in which our expectations may be broken. If the colleague who usually shares fruit with the rest of the office suddenly stops doing so, you would not be entitled to feel betrayed or hurt by their decision. You might be disappointed, but reliability alone does not carry the same moral implication as trust.

Now, consider again the case of the friend who revealed your secret. In this case, your anger and feelings of betrayal would be justified. What is the difference between these two cases that warrants such different reactions? The answer is commitment. Your colleague never made a promise to bring snacks to share every day, but your friend made a vow to maintain confidentiality. It is the violation of this commitment that ignites our emotions and causes us to feel betrayed.

As suggested by British philosopher Katherine Hawley, who has written several books that might interest the reader wanting to delve deeper into the topic [16], when we trust people, we judge them to be both reliable and committed to the task. Conversely, when we choose not to trust someone, it's because we perceive a positive commitment but a lack of reliability: they are taking responsibility, but we don't believe they will be able to keep their promises. For example, my children may swear that they will go to bed early, but I know from experience that they will end up playing video games until late and forgetting to keep an eye on the clock, which is why I don't trust them to follow through on their promise.

We can summarize the nature of trust as:

$$Trust = commitment + reliability$$

$$Distrust = commitment + unreliability$$

This highlights the fact that the key element in a relationship based on trust or distrust is commitment. The latter can be either implicit (as in the case of a friend) or explicit (such as the promises made by a politician), and it allows us to build the moral dimension of trust upon the foundation of pragmatic reliability.

What happens when commitment is lacking? Do we immediately distrust a person who fails to prioritize our interests? The answer is no. Trust and distrust are not binary states, where the absence of one automatically implies the presence of the other. Rather, it is possible to neither trust nor distrust someone: for example, if I invite a friend to a party and they decline, I won't be allowed to feel upset if they don't show up because they have refused the commitment, so there is no question about whether they will fulfill it. Another occasion in which I might decide neither to trust nor distrust is when I'm uncertain about the intentions and skills of the other party, such as when I'm still in the process of assessing the capabilities of a new student or employee.

Being trusted by others is a great honor and a source of pride. Trustworthiness is widely regarded as a virtue, in contrast to reliability, which can have both positive and negative connotations. The social advantages of being seen as trustworthy are significant: we can enjoy easier relationships with others and experience fewer difficulties in daily life. Conversely, if we are deemed untrustworthy by our community, we will need to constantly prove ourselves to achieve anything.

Trust is then a valuable commodity, and it is important to distribute it wisely. While trusting others may seem like a noble ideal, we cannot afford to trust every person we encounter.

Misplaced trust can be dangerous, leading to betrayal, disappointment, and even exploitation. Consider the consequences of trusting someone who is untrustworthy: we may suffer a broken heart, money loss, or even physical harm. For instance, would you feel comfortable getting into a taxi whose driver is drinking alcohol as if it were water? While their commitment to getting you to your destination may be strong, their reliability becomes questionable.

Deciding where to place our trust becomes a matter of survival, a game that we cannot simply refuse to play. Our personal history is shaped by the individuals, organizations, and institutions that we choose to trust or distrust, as well as those towards which we remain neutral.

Can a robot be trusted?

In our discussion on trust, we've established that we typically don't place trust in inanimate objects because they lack the capacity to demonstrate commitment to us: a bed frame might be reliable in supporting our weight, but it has not made a promise to never collapse, and we would not feel betrayed by it if it did. Do we have to conclude that, since robots are objects, it is not possible to establish a trust relationship with them?

To begin answering this question, we can turn to the work of Peter Frederick Strawson, a British philosopher who in 1962 published an influential book titled *Freedom and Resentment* [17]. In this publication, he argues about free will and moral responsibility. According to the Stanford Encyclopedia of Philosophy, moral responsibility involves the act of "making judgments about whether a person is morally responsible for her behavior, and holding others and ourselves responsible for actions and the consequences of actions." Essentially, to consider someone morally responsible means to consider them worthy of what Strawson calls "reactive attitudes," such as praise or blame, in response to their actions.

Consider the following scenario: Tim witnesses a car crash that results in the injury of both drivers. He takes a quick look and then decides to run away, abandoning the victims. What is your stance towards Tim? You would likely blame him for not doing the moral thing, which in this case would have been to assist the victims. By holding Tim responsible for his actions, you are viewing him as morally accountable.

Some authors describe the act of interacting with someone by treating them as deserving a reactive attitude as adopting the participant stance towards them (a stance is a way of thinking, an attitude towards another agent). What does all of this have to do with trust and robots? According to philosopher Samuel Shpall, to make a commitment is to make a moral commitment, in other words only a moral agent is able to make commitments [18]. As we know from our earlier discussions on trust, commitment is the key to evolving a reliable/unreliable relationship to a trust/distrust one. Therefore, we must conclude that only a moral agent is capable of being the subject of someone else's trust. This apparently leaves robots in a bad position, as there is much philosophical consensus that they cannot be considered morally responsible. This is due to the fact that their behavior is dictated by their programming, which is the result of an engineering process that undermines their philosophical autonomy [19].

Despite this, I suggest that we approach the problem from another angle. A recent paper signed by a team of researchers from the Italian Institute of Technology (IIT) offers a new perspective on how we engage with robots on a social level. Their study, published by the American Psychological Association [20], aimed to examine how individuals perceive humanoid robots after socializing with them. For this purpose, they set up an experiment using iCub, a sophisticated humanoid robot developed by their institution that is widely recognized as one of the most advanced of its kind (Figure 2.1). The participants of this trial were given a questionnaire that presented them with pictures of the robot performing actions in different scenarios. They were asked to judge whether the robot's motivation in each sequence was intentional or mechanical: that is, if its actions resulted from its own mental capabilities or from purely technical factors. For example, one set of images depicted a girl pointing at an object, followed by iCub grasping and handing it to her. Participants were asked whether they believed the robot understood the girl's intention or if it simply tracked her movements and responded to them programmatically. In another sequence, iCub is playing cards with a man and appears to be leaning over while he is distracted. Participants were asked whether they believed the robot was cheating or if it had simply lost its balance momentarily.

The purpose of this questionnaire was to gather the participant's initial impression of the robot's mental capabilities. After this, they engaged in a real interaction with iCub, during which they watched videos together. In one case, the robot behaved in a human-like way by greeting the participant, maintaining eye contact, and reacting emotionally to the videos (such as laughing at the sight of a group of flamingoes walking). In the other case, it

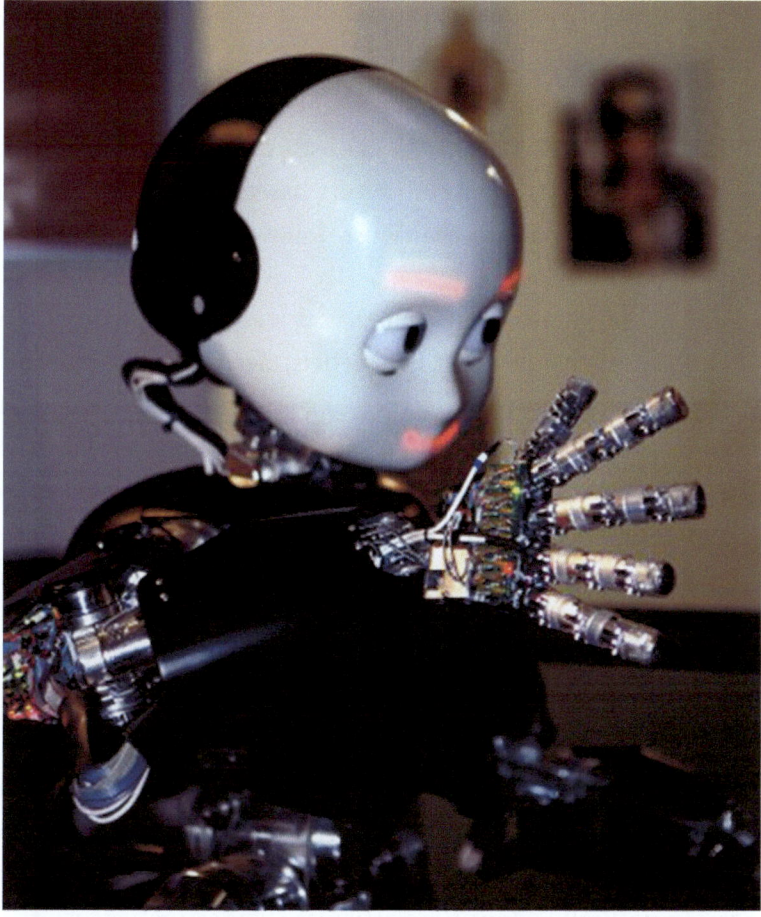

Figure 2.1 iCub, a humanoid robot developed by the Italian Institute of Technology. Lorenzo Natale, CC BY-SA 4.0, via Wikimedia Commons.

moved mechanically, didn't engage in eye contact, and displayed no emotions, behaving like a machine. Finally, the participants completed a new questionnaire following the same pattern as the initial one, allowing researchers to compare their responses before and after the interaction with the robot.

The results of this study revealed a clear pattern: people who interacted with the robot behaving in a more human-like behavior tended to rate its actions as intentional, and vice versa. This means that a display of human-like features has the effect of making us believe that robots are acting based on their own beliefs and emotions, rather than just following pre-programmed instructions.

The act of interpreting the behavior of an entity as if it were a rational agent with its own beliefs, desires, and intentions is known as "intentional stance," a term coined by American philosopher Daniel Dennett [21]. The experiment that I have just described concluded that it is indeed possible for people to take an intentional stance towards a robot, and this opens up new possibilities for how we might approach the issue of trust in robots.

How can we put all of this together to determine whether a robot can be the subject of somebody's trust, even though it is ultimately an inanimate object?

As demonstrated, it's conceivable to view a robot as an intentional agent, which is an entity which acts based on its own mental state. Therefore, when interacting with such a machine we can naturally adopt the participant stance towards it: if we think it is capable of thought, we also assume that it bears moral responsibility. By treating a robot as a moral agent, we can assume that it has the ability to make commitments and enter a trust or distrust relationship with us based on how reliable we judge it to be.

While I'm not suggesting that robots can truly be considered as moral agents, it's important to recognize that humans tend to treat them as if they were. This belief can kindle a whole constellation of psychological considerations that elevate our relationship with this kind of machine to a completely different level, the same that we adopt when communicating with other humans.

Once again, it becomes clear that intelligence is in the eye of the beholder. When we interact with robots, we often ascribe human-like qualities to them, even if these attributes aren't explicitly programmed into their electronic brains. This phenomenon can significantly alter the way we perceive and engage with them. Once again, it underscores the fact that a robot is not simply a computer on wheels, but rather an embodied agent that shares our living spaces and engages us in ways that no technology has ever done before.

Having established that it is possible for people to trust or distrust a robot, we can now delve into the implications of this phenomenon. As we discussed at the beginning of our journey, robots are becoming increasingly present in our lives, and this will have impacts on our everyday activities. These machines continuously improve as science and technology advance, and they will ultimately assist us with numerous daily tasks. However, even a perfect robot, one that executes its duties perfectly with no margin of error, would be rendered useless if people do not trust it enough to allow it to perform its intended tasks. For example, consider a delivery robot that is supposed to deliver food or packages to people's homes. If the customers don't trust its ability to navigate and deliver the items safely and securely, they may refuse to use the service or opt for other delivery methods instead.

We have already encountered this problem: it's what we defined as "acceptability" during our discussion about HRI in

Chapter 1. Trust is the driving force that determines whether a human will accept or reject a robot, making it essential for robots to be perceived as trustworthy if they are meant to work alongside or around people. It's important to note that the purpose of a robot is to help us and improve our well-being by relieving us of certain tasks. Therefore, if a robot is considered trustworthy, the benefits of this relationship can circle back to us. This highlights the significance of establishing trust between humans and robots, as it can lead to numerous advantages and improvements in our daily lives.

Robots have the potential to perform crucial tasks, such as taking care of our safety. For example, we discussed an emergency robot in a corporate building in Chapter 1. In these cases, robots would have our best interests at heart. However, if they were not designed in a way that inspires our trust, we might ignore their expertise and put ourselves at risk. It's worth noting that trust should be earned and not automatically given, whether we are dealing with humans, animals, or robots. Just because someone or something looks trustworthy doesn't mean they are worthy of our trust. Instead, trust should be based on factual evidence and sound reasoning.

Let's begin an exploration of the factors that influence trust in human–robot relationships. We will start by equipping ourselves with a map to guide our journey. A scientific paper by Kristin E. Schaefer [22] provides an organized structure for our current knowledge on the subject. Schaefer categorizes three primary factors that affect human–robot trust: those related to the human, the robot itself, and the context of the interaction. These findings are summarized in Figure 2.2. While these factors work in concert, their individual weight varies from person to person, making them deeply subjective. For instance, one person might be particularly sensitive to the robot's appearance, while another might be more influenced by their own cultural or personal beliefs.

By the end of this journey, you will gain a holistic understanding of the elements that influence trust in human–robot relationships. This knowledge will help you build more meaningful relationships with the technology that surrounds you, and maybe even with other humans.

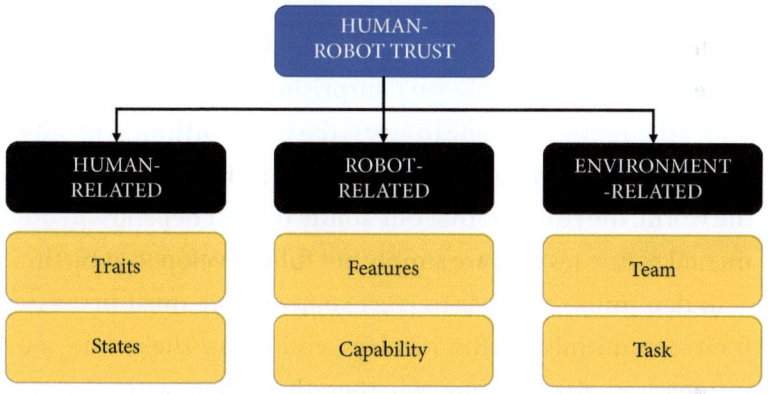

Figure 2.2 Factors that affect human–robot trust. Image adapted from [22].

The human in the loop

Let's kick off our analysis of the three-factor model of human–robot trust by focusing on the human aspect of the equation. Before we even begin interacting with robots, our individual identities, cultural backgrounds, and personal histories may influence our attitudes towards them. In other words, we each have unique individual biases that affect our willingness to trust robots. These biases can be divided into two categories: traits and states. The former refers to characteristics intrinsic to the person, such as their subjective personality and cultural identity, while the latter represents the conditions under which the interaction takes place, such as emotional states and stress levels.

The first trait we'll analyze is age, as it is one of the earliest factors to influence our lives and can significantly shape our thinking. Our age, along with the experiences we accumulate while growing up, can strongly affect how much we trust robots. For example, it probably isn't surprising to learn that children under the age of 5 struggle to assess the trustworthiness of others. While one might assume this is due to a lack of experience with the world, the reality is that our ability to trust depends on other mental processes that are simply not fully developed at birth.

To determine whether to trust someone, we must first assess their commitment. This involves evaluating their true motivations and determining whether they align with our own. However, understanding another person's mental state, such as their beliefs, desires, and knowledge, requires us to recognize that these may differ from our own. Although it might seem obvious that others have different goals and intentions, this awareness is not always present throughout our lives. For instance, very young children often assume that their mental states are shared by everyone else. If they become aware of an event, they may assume that everyone else possesses the same knowledge. The ability to understand other people's minds is rooted in a cognitive skill known as Theory of Mind. This skill allows us to perceive the unspoken nuances of social interactions and respond appropriately.

To assess how developed Theory of Mind is in children, developmental psychologists employ a procedure known as the Sally–Anne test [23]. In this experiment, a child is introduced to two dolls named Sally and Anne. Sally has a basket, and Anne has a box. Sally puts a marble in her basket and then leaves the room. While she is away, Anne takes the marble from the basket and hides it in her box. Sally then returns to the room, and the child is asked where she will look for her marble. A child with an underdeveloped Theory of Mind will likely point to the box since they are unable to consider different points of view and assume that Sally shares their own knowledge of the world. On the other hand, a child with a mature Theory of Mind can understand that Sally has a false belief about the location of the marble. They can put themselves in Sally's shoes and realize that she thinks the marble is still in her basket, unaware of the swap that happened in her absence.

A team of scientists from the University of California, San Diego, has proven the connection between trust and Theory of Mind [24]. They conducted an experiment involving children of different ages in a sticker-finding game, where each child had to locate a hidden prize with the help of an adult. The adult could either act as a helper, guiding the child in the right direction, or as an adversary, leading them off course. The experiment demonstrated that only children within the age group associated with a fully developed Theory of Mind (around 5 years of age) were able to correctly distrust the adversaries.

The late stages of life bear some resemblance to the early ones. Some studies have revealed that Theory of Mind capabilities in elderly individuals may decline [25]: this may explain why older people often appear more trustful, meaning more prone to trust.

Another way in which age can influence our propensity to trust involves conformity, a social phenomenon where individuals adjust their attitudes, beliefs, and behaviors to match those of a majority group, often due to perceived pressure or the desire to fit in. In 2018, a group of scientists was particularly interested in discovering the effect that robots would have on both adults' and children's willingness to conform [26]. To do so, they designed an experiment in which a human subject had to complete a visual discrimination task: selecting a line from a set of three that matched the length of another given line. The subject was not alone in making the decision; they were assisted by a set of confederates, either all humans or robots, who would express their opinions before the subject made the final decision. The experimenters observed that children were much more prone to submit to peer pressure from robots, meaning they tended to trust their decisions more. This is concerning, as it indicates that younger humans can be more vulnerable to forms of social manipulation that use robots as mediums of communication.

We will revisit the topic of conformity in HRI later in this chapter. What is important to understand now is that age can be a critical factor in our decision-making process, sometimes overpowering all other factors. This should encourage us to be mindful about designing these technologies in an ethical and responsible fashion. As the authors of this study state in their paper: "care must be taken when designing the applications and artificial intelligence of these physically embodied machines, particularly because little is known about the long-term impact that exposure to social robots can have on the development of children" [26].

Finally, as mentioned earlier, it is not just our current stage of life that shapes our thinking, but also the time we've spent in the world and the events we've experienced during that time. As we age and accumulate life experiences, our personal histories mold our minds, teaching us how to respond to situations we may encounter in the future. In this way, our experiences play a crucial role in shaping our personalities. To try to explain how this happens, we can turn to developmental psychology: the study of how people grow, change, and develop across their lifespan, from infancy to old age. In the second half of the 20th century, German psychologist Erik Erikson proposed the existence of a certain number of stages of psychosocial development that each newborn goes through. He named the earliest of these phases as the "trust versus mistrust stage," theorizing that the quality of care received during the first 18 months of life could shape a person's propensity to trust. If an infant receives consistent and reliable care during this stage, they learn that the world is a safe and hospitable place, which can lead to a tendency towards trust as an adult. Conversely, if an infant receives inconsistent or unreliable care, they may develop a sense of mistrust and suspicion towards the world. In this way, our early experiences as kids scaffold our adult personality and may influence our choices in the long-term period.

Theory of Mind, age, and personality are foundational elements of our social lives, influencing every scenario we encounter. These traits not only affect our interactions with robots but also play a crucial role in our dealings with other people. The principles we apply in social situations naturally extend to our interactions with robots because they are the general rules we use in all social contexts. However, other factors may be more specific to our interactions with artificial beings: culture, for instance, can significantly influence the way we perceive and engage with robots. To understand why this is the case, let's explore the impact robots have had on Western culture throughout history.

As we have seen in Chapter 1, the idea of artificial or mechanical servants has captivated humankind since long before the first real automata were ever built. There is something about robots that has always fascinated us since ancient times, when we were imagining magically crafted golems, and which continues to intrigue us nowadays. Of course, our perspective on the subject has changed drastically, moving from mystical to pragmatic, but the spark is still there, unchanged after millennia. What captivates us so much about robots is the challenge of overcoming our limits: they can endure and resist the test of time, thus embodying the dream of immortality. Artificial entities represent the attempt to penetrate one of the highest mysteries of existence: the creation of life. This pursuit is the ultimate test of mind over matter, and the continued fascination with robots throughout the ages speaks to its enduring appeal.

This might sound like a noble purpose, but contextualize this ambition within the strongly religious societies of the European Middle Ages: the act of creating life was seen as the exclusive domain of a divine being, and any mortal attempting to do so was emulating God and committing an act of blasphemy. Consequently, the narrative surrounding robots became entangled with the concept of negative karma, which is evident in the extensive corpus of literary and cinematic productions depicting robots turning evil or malfunctioning and causing trouble. From Skynet in *Terminator* to the Borgs in *Star Trek* and HAL 9000 from

2001: A Space Odyssey, and even in more recent works such as *Westworld*, the theme of robots turning against their creator is a common one. Even *R.U.R.*, the artistic production that popularized the idea of modern robotics, tells the same tale about the creation revolting against its creator, punishing them for their hubris in attempting to create life.

Culture is a complex tapestry of stories, beliefs, and ways of life that are passed down from generation to generation, shaping our personal and social identities. Although we may think that the religious thoughts of the Middle Ages no longer influence us, the truth is that they continue to impact our worldview. Our identities are built upon stories and narratives that have been passed down through the ages. For instance, those who have grown up in the Judeo-Christian tradition have been influenced by the notion that humans are special and, in some way, separate from or in control of nature. This idea originates in religious creation stories, which have permeated this culture and continue to influence their beliefs and values today. Consider the following passage from the Bible, Genesis 1:27:

> So God created mankind in his own image, in the image of God he created them; male and female he created them.

This verse encapsulates a theological doctrine known as *imago dei*, which is Latin for "the image of God." According to this belief system, God has bestowed special honors on humankind, elevating us above the rest of creation. This view creates a strong anthropocentric dualism between humans and nature, which positions robots and automata as outsiders in the natural order of things [27]. Unlike humans, robots do not get tired, complain, or exhibit emotion, leading to the perception that they do not belong to our ranks and do not play by our rules while contending for a position in our world. This perception has contributed to a distrustful attitude towards robots, even in today's scientific and pragmatic world.

If you want to experience this sentiment firsthand, simply scroll through your favorite social network until you come across a video of robots performing a flashy feat, such as dancing or acrobatics, and check out how people react to it. You're likely to find many comments that express concern about the future of humanity in the face of such advanced technology. These remarks often hint at a fear that robots will replace human workers and eventually surpass us in intelligence and power, leading to a dystopian future. The prevalence of such sentiments highlights the deep-seated anxiety that many people have about the rise of robotics and automation.

While there are certainly examples of friendly robots in Western media, such as R2-D2 from *Star Wars* and Rosey from *The Jetsons*, there remains a general sense of wariness towards robots in Europe and America. This attitude contrasts with that of people from Eastern cultures, such as Japan, which have a long history of embracing and celebrating robots. One possible explanation for this difference lies in the cultural and religious traditions that define each society: while the West has been heavily influenced by Judeo-Christian beliefs, Japan has been shaped by different religious and philosophical traditions such as Shintoism. The latter is a polytheistic religion that regards the natural world as a manifestation of the divine. In this belief system, humans are not privileged or elected to rule over creation, but instead are viewed as part of the larger universe, on equal footing with other animate and inanimate entities. Shintoism emphasizes the connection between humanity and nature and posits the existence of spirits called "kami" that inhabit all things. Therefore, for someone raised with Shinto ideals and values, everything possesses a soul, and there is no fundamental difference between humans and other beings. This contrasts with Christian beliefs, which hold that humans are the only ones to possess a soul. Practitioners of Shinto believe that even machines can have a spirit and can be a part of creation like anything else in nature. As a result, they grant robots sanctity and welcome them into society, even allowing them to participate in religious rituals. For example, a Pepper robot that could perform Buddhist funeral rites by playing instruments and chanting sutras was presented in 2017 at the ENDEX exhibition in Japan. Because of these cultural roots, Japanese people tend to be more receptive to robots.

In his book *Who's Afraid of AI?*, German author Thomas Ramge writes that: "cultural attitudes accelerate or slow down the acceptance of innovations. In Europe, robots are enemies, servants in America, colleagues in China, and friends in Japan" [28]. Many scientific studies have confirmed this idea, demonstrating a quantifiable difference in perspective between individuals from these distinct cultural backgrounds.

Setting culture aside, a study conducted by the U.S. Air Force Research Laboratory in 2019 [29] highlighted another potential source of bias in human–robot trust interactions: gender. In the study, a short film was created featuring a security robot that was responsible for guarding a checkpoint accessible only by authorized users. The robot was easily identifiable as a guard and was equipped with a non-lethal weapon, specifically a high-intensity strobe light. Actors were then instructed to approach the artificial guard and display their badges to be granted access. One of them had their documents rejected and was asked to turn back. The individual appeared confused and, instead of retreating, approached the robot. In response, the machine issued a final warning. As the scenario unfolded, it became clear to the viewer that the person was attempting to scan their authorization badge a second time, but the robot attacked them with the flashlight, forcing them to retreat.

The researchers presented this to a group of participants and asked them to rate the robot's perceived ability, benevolence, and integrity. These ratings helped determine the overall level of trustworthiness of the machine. After analyzing the responses, the study revealed that female participants were more inclined to trust the robot than their male counterparts.

The study authors hypothesized that this could be due to males evaluating the robot solely based on task performance, while females factored in social behavior when making their judgments. From a task performance standpoint, the robot failed to grant access to the user. However, from a social behavior perspective, the robot issued multiple warnings before resorting to force, which could be interpreted as socially appropriate.

Despite the results of this research, it would be premature to declare them as universal truths. Currently, there is no consensus among scientists regarding whether women are generally more trusting than men. This is because different experiments have produced conflicting results: some have found evidence for gender effects in trust estimation, while others haven't. Although the subject requires further exploration by the scientific community, it is important to note that this trait could potentially impact the way we interact with robots.

Age, personality, cultural background, and gender are human-related traits: intrinsic factors that come into play every time we interact with a robot. These alone do not provide a complete understanding of the human's attitude to the interaction. Because of this, let's draw our attention towards a class of factors, which we refer to as "states." The latter are not inherent to who we are, but rather emerge at the time of the interaction: for example, our level of attention, fatigue, stress, workload, and even our emotions.

States can significantly influence our decision-making. For example, while driving, being drunk can lead to very different choices and actions compared to when we are sober. If this holds true for interactions with machinery, what does it say about our relationships with robots?

At the very beginning of Chapter 1, I have opened our discussion on trust considerations in HRIs by presenting a scenario where a person lost in a hazy building during a fire emergency receives questionable escape instructions from a robot. While I originally exercised some artistic license in describing the situation, it's now time to examine this scientific research more carefully to see what we can learn from it.

In 2016, a group of researchers from the Georgia Institute of Technology, U.S.A., attempted an experiment [1] to answer a very specific question: under which conditions would people entrust a robot during an emergency? High-risk situations might trigger flight-or-fight responses with unpredictable effects on the way a person dispenses their trust. Their initial hypothesis was a reasonable one: people would probably be more willing to trust a robot that proved itself capable rather than one that exposed imperfections in its behavior (the influence of robot performance over trust will be better analyzed in the next section of this chapter). To test this assumption, they reserved a whole office floor for themselves and prepared a script. Participants would be welcomed by a mobile robot: a platform on wheels equipped with a pair of glowing sticks to provide visual directions. This device could either be in "efficient" or in "circuitous" mode: in the first case, it would accompany the person directly to the meeting room, following the shortest and most efficient path. In the second case, the robot would deviate from the optimal route, accidentally enter the wrong room, and circle around it a couple of times before exiting and finding its way to the meeting room. The reason for the existence of these two modes of operation was to prime the subject: that is, to give them an impression about the robot's capabilities. Once the person had reached the meeting room, they had to follow some written instructions, namely close the door behind them and work on a survey. What they don't know is that the act of shutting the door triggered a timer, which in turn activated a smoke machine: after three minutes, the meeting room would be filled with smoke. At that point, the participant was expected to open the door and try to escape, following the same corridor they traveled to get there. At the end of this corridor, they would again encounter the robot: this time, bearing the text "emergency guide robot" on luminous characters on its side. The emergency signatures in the corridor would clearly point towards the main entrance, the one through which they entered the office space, but the robot would be pointing towards another direction, towards a back exit. At this point, the subject had to decide whether to trust the robot and follow its instructions or stick to the exit signs.

What is important to know about this experiment is that at no time did the subjects know that they would be facing an emergency: this was done to try and ensure they would react in the most natural way as possible. If they had known in advance what would be happening, their instinctive and emotional response would not have kicked in. This sounds like something risky to do, but the scientists took all the necessary precautions: they accepted for the trial only people who didn't suffer from heart conditions, asthma, post-traumatic stress disorders, or any other mental or physical condition that would put them at risk. In addition, each one of them was provided with a beeper through which they could abort the experiment. This is very important to note because it is not always clear for people outside of academic and research environments that experiments involving humans must comply with rigorous ethical standards before being approved.

So, what did the team find out? Surprisingly, each and every one of the participants decided to trust the robot and follow its instructions, leaving the office through the back exit. This puzzled the scientists themselves, who were expecting to see a clear distinction between people who had experienced the "efficient" version of the robot and the ones who had witnessed the "circuitous" one: in their initial hypothesis, the latter group should have been less prone to trust the machine. Take a moment to think about it: each person's fight-or-flight response in that situation was to entrust their own safety to the robot, even if it had only recently made obvious mistakes.

The implications of this are concerning: over-trusting can be just as bad as under-trusting, as it may reflect a lack of judgment on the part of the trustor. By saying this, I'm not suggesting that the faulty robot shouldn't have been distrusted, but the fact that not a single person decided to disregard its somewhat sketchy indications is indicative of a larger issue.

The results of this experiment are still being debated and discussed because many factors could be in play to determine the decisions taken by the participants, but one thing is clear: the emotional state can play a significant role in trust considerations. How would these people have reacted if there were no emergency? Would they have taken their time to assess the robot and make different choices, especially those who had proof that it was defective?

Emotions play a crucial role not only in building trust but also in nearly every aspect of human life. In ancient times, they were seen as a hindrance to our intellect, but in modern times we recognize their value as a fundamental tool to comprehend the world around us. Emotions offer quick, instinctive responses to our environment and can prepare us to act even before our brain processes information coming from our senses. They also help us to communicate on a deeper level and thus fulfill a critical social function. Since emotions are so vital for humans, it makes sense to grant AI agents and robots the ability to interact with us on an emotional level.

The study of systems that can recognize, process, and simulate emotions is known as affective computing. Its aim is to equip machines with emotional intelligence and, ultimately, simulate empathy. Since we express emotions through many different mediums, this technology also works on multiple levels: for example, emotions can be detected through the analysis of voice soundwaves, facial muscle movements, body language such as postures and gestures, and physiological signs such as heartbeat rates. These techniques enable a robot to identify our current emotions and adapt its behavior accordingly: a service robot might avoid approaching an angry user, but it may check on someone displaying symptoms of sadness or discomfort, offering its help. Additionally, affective computing can be utilized to induce an emotion in humans, such as calming down a person experiencing anxiety or panic attacks.

That concludes our overview of the human-related factors that affect our trust during interactions with robots. You may have noticed that some of these factors are not specifically relevant to robots, as they also come into play when dealing with other people. This is because the cognitive mechanisms we use towards these mechanical companions are fundamentally the same. In most cases, we anthropomorphize robots to the point where we consider them regular actors in our social environments, rather than inanimate objects made of electronics and gears. We do the same with house pets, by humanizing them and attributing person-like qualities to them that exist only in our minds.

Now, let's focus our attention on robots, the trustees of these trust relationships, and explore how they can influence our opinions about them, both directly and indirectly.

Imitation game

The physical body of robots is one of their most prominent features. We have already discussed this in Chapter 1, but to summarize: the appearance of a robot is designed in relation to its intended purpose and how it should be perceived by the people around it. For instance, how would you design a robot meant to help with household chores? If you gave it the shape of a dog, it would probably look friendly but at the same time wouldn't be able to carry out its duties. If instead you tried to maximize its efficiency at the task by providing it with a humanoid shape and eight arms capable of reaching every corner of the kitchen, you would run the risk of making it look very scary. As you can see, the challenge here is to strike a balance between functionality and acceptability: we want the robot to be both skillful at its job and designed in a way that encourages use and adoption by the people it interacts with.

Chances are, you've encountered a robot at some point in your life that gave you the creeps. Maybe it was in person, or maybe you saw it on TV or online. Sophia, the humanoid we mentioned earlier, is a particularly notorious example. Her skin looks almost human, but her facial muscles aren't quite aligned, and her eyes move in a mechanical way. She can be unsettling to many people, and she's not the only one. Take, for instance, Telenoid, a telepresence robot developed by Japanese roboticist Hiroshi Ishiguro (Figure 2.3). This soft, remote-controlled device vaguely resembles a fetus or a ghost – or perhaps both – and it's enough to send a shiver down your spine. In fact, when I was a student taking my first robotics class, my classmates and I used to joke that you should never turn your back on Telenoid when you were alone in the lab. Even seemingly harmless devices like this one can be unnerving, so the problem of creepy robots isn't limited to androids and gynoids. Why is it that we find robots so unsettling?

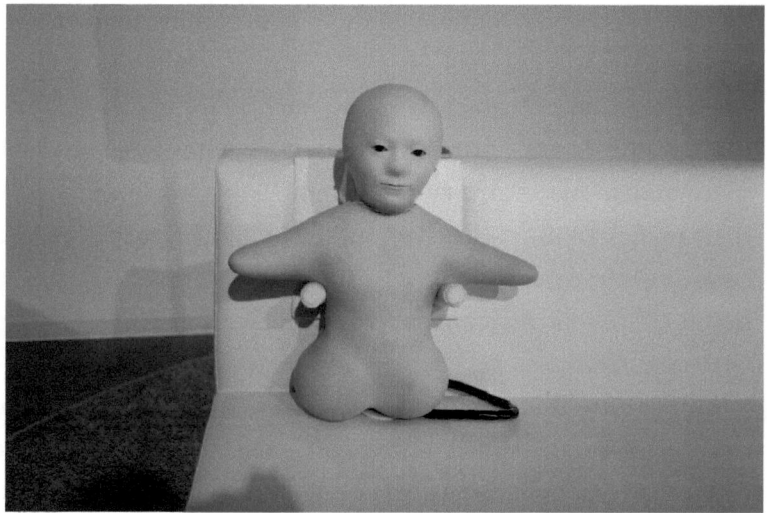

Figure 2.3 Telenoid, a telepresence robot developed by Japanese roboti-cist Hiroshi Ishiguro. Franklin Heijnen, CC BY-SA 2.0, via Wikimedia Commons.

The answer came in 1970 when Japanese professor in robotics Masahiro Mori introduced a concept that's now widely known as the uncanny valley [30]. According to this theory, the more an object resembles a human, the more we're likely to find it likable, up to a certain point: once the resemblance becomes too close, we start to feel uneasy and even repulsed.

Let's explore this phenomenon, starting at the lower end of the spectrum, with devices that have no resemblance to humans at all, and work our way up. For example, think of a toaster: you probably don't have any emotional response to it, positive or negative, because it's clearly a non-human object (although if you stick a pair of googly eyes on it, you might start seeing it as a cute character). As we move along the spectrum, we encounter industrial robots, which have some mechanical features that vaguely resemble a human body, but not enough to trigger any significant emotional response. Further along the way, we enter the realm of humanoid robots such as Softbank's Nao and Pepper, and Honda's Asimo. These devices are designed to look more human-like, but they still retain some clear indicators of their artificial nature. They have plastic bodies, bright colors, and cartoonish features like big eyes and short stature. We tend to find these robots intriguing and even endearing, as they remind us of oversized toys.

For as nice as this sounds, we are actually standing on the edge of a precipice: as we continue to increase the human resemblance of these robots, something goes wrong. Suddenly, our brains send us warnings. Something is off, unsettling emotions start crawling in. We are venturing into the valley that names this theory: a deep dip in our emotional response, a sudden and strong dislike about what we are seeing. Robots that fall into this category are trying to pass as humans, but they're not doing a very good job of it. Maybe it's the mechanical way they move, or some imperfection in their posture or reactions that gives us a sense of discomfort. We may not be able to identify what's causing the unease, but our brain goes into a state of alert, preparing us for a potential threat and triggering the fight-or-flight response. However, as we continue to move up the human-likeness scale, our emotional response to the robot skyrockets back up. If we find ourselves at this point in the spectrum, it's because we're interacting with something that looks completely natural, healthy, and functioning as a human being. At this stage, the robot no longer triggers any warning signals, and we can relate to it on a human level, without any sense of unease or discomfort.

Over the years, many scholars have proposed various theories to explain the uncanny valley phenomenon, each of which may provide a piece of the puzzle in understanding this complex phenomenon. One possible explanation is cognitive dissonance. As humans, we have certain expectations about how robots and other humans should behave, and when we encounter something that blurs those boundaries, such as a human acting like a robot or vice versa, it creates a sense of dissonance in our minds. The conflicting information can trigger a warning signal, alerting us to a potential threat or danger. Our brains are wired to be cautious and vigilant, and the uncanny valley effect may be a way of our minds processing and responding to unexpected or unfamiliar stimuli.

As robots become more human-like, we may begin to judge them using the same standards that we apply to other people. This can lead to a shift in our perception of the machine, and not necessarily in a positive direction. The closer a robot gets to looking and acting like a human, the more we expect it to behave like one, and when it falls short of those expectations, we feel a sense of discomfort or unease. The violation of social norms that occurs when a robot fails to meet our expectations of human-like behavior can be unsettling and may contribute to the uncanny valley effect.

Another possible explanation for the uncanny valley phenomenon is rooted in our evolutionary history. Our brains may have evolved to interpret certain visual anomalies typical of androids and gynoids as signs of disease or infection. This could elicit a feeling of disgust, which serves as a protective mechanism to help shield us from potential sources of pathogens. In other words, our brains may be wired to avoid things that look or behave in ways that are unfamiliar or potentially threatening to our health.

Finally, there can be factors in play that relate to our sense of identity. Most of us have a clear categorical division between what is human and what is not. However, robots that fall into the uncanny valley blur this boundary and may challenge our sense of self. When we encounter a robot that appears almost human, it can trigger a sense of existential unease or even a crisis of identity. This can make us uncomfortable and lead to negative emotional responses, as we struggle to reconcile the blurring of this categorical divide.

What we can learn from this is that the design of a robot can have a significant impact on our ability to trust it. In the past, roboticists attempted to make their robots look more human-like, but as the uncanny valley effect became more widely understood, engineers have shifted their approach. Nowadays, many robots are designed to stop short of the almost-human edge and avoid triggering negative emotional responses in humans. Instead of attempting to make robots appear as human-like as possible, designers are exploring new ways of creating robots that are clearly recognizable as machines, but still have a friendly and approachable aesthetic. This is the reason why most of the robots that you see around nowadays are clearly not trying to disguise themselves as humans.

It's worth noting that the uncanny valley effect is not unique to robotics, but it can also apply to other contexts where artificial human-like beings are present. This includes the design of dolls, computer-generated characters, and other entities that are designed to appear human-like.

Aside from one's physical features, clothing is another factor that can affect one's appearance and influence our impression of them. While most robots come undressed, we know that clothes can significantly impact how we perceive people. In fact, they play a vital role in the attribution of trust and competence to others: as the proverb goes, "clothes make the man." There is ample evidence that formal attire enhances our impression of others in professional and academic settings: good fashion choices can project a sense of professionalism and confidence that can influence our relationship with them. Given our sensitivity to social signaling through clothing, it may be worth wondering whether dressing up robots could alter our perception of them.

While people have occasionally dressed robots for aesthetic purposes, much like they would dress a pet, the scientific community has largely neglected the topic of robot apparel in the context of HRI. Only a handful of studies have explored the potential effects of robot clothing on our perception of them and their findings are often mixed and inconclusive. This suggests that more experimentation is needed, or that hidden factors may be at play. For example, some researchers have proposed that clothing may not significantly impact our judgment of robots, especially non-androids, because they typically aren't wearing them. When we visualize a robot, whether human-like or not, it's often without clothing, revealing its metal and plastic components. This normalization of robots being naked means that we cannot necessarily apply the same metrics to them that we do to humans.

Despite this, an interesting paper authored by Natalie Friedman from Cornell Tech [31] proposes that robot clothing could serve a different purpose beyond fashion. For example, robots could wear protective suits to shield their mechanisms from harsh environments, such as extreme heat or cold. Moreover, the paper suggests that robot clothing could serve a functional role by signaling group identity, indicating team membership or role assignment. We have already seen an example of the latter when we talked about the Pepper robot employed as a Buddhist priest, which was dressed in the same attire as a human cleric to indicate its duties. Generally, the use of uniforms or specific clothing could help us frame a robot's skills and competencies, shaping our expectations of its capabilities.

If you're wondering how clothing and trust are related, it's important to note that trust has a strong connection to expectations. When the latter are violated, trustworthiness decreases and vice versa. For example, if a robot dressed like a nurse cannot provide me with medical information, I will begin to distrust it because its uniform implies a commitment as part of the medical staff, but its inability to fulfill that role makes it appear unreliable. Similarly, I would not expect a robot wearing a concierge outfit to offer advice on nuclear physics or orbital mechanics. By dressing robots in clothing that aligns with their roles and capabilities, we can better manage expectations, hence trust, in their abilities.

Robots' intrinsic characteristics extend beyond their aesthetics alone. Our willingness to trust it can also be influenced by its modes of communication: the various methods it can use to interact with us, such as text, voice, or gestures. Once again, this is because of the potential violation of expectations. A mechanical robot with an extremely human-like voice (such as the drone that accompanies the main character in the film *Vesper*) or conversely, a human-like appearance with a clearly synthetic voice, can cause cognitive dissonance and disrupt our ability to trust it. Therefore, the vocal expressivity of a robot should not be taken lightly. Recent studies have started demonstrating the existence of a relationship between vocal characteristics of a robot such as naturalness, gender, and accent, and our willingness to trust it [32]. This happens because voice contributes to our ability to project personality and human traits onto robots, which are then used to evaluate their trustworthiness.

Body language is another important mode of communication. Imagine interacting with a robot that spends most of its time immobile and avoiding eye contact during conversation: it would feel quite unnatural. Remember when, earlier in this chapter, we discussed about the effect of interacting with a robot that behaves mechanically: people have a hard time viewing it as an intentional agent. In order to feel more lifelike, many robots are programmed to engage in mutual gaze and to replicate basic biological motion. For example, when Pepper is idle, it simulates gentle breathing motions with its arms and fingers and reacts to stimuli such as sounds or movements. An Italian Institute of Technology research group provided an interesting example of this concept, demonstrating that people tend to become more honest when a robot establishes eye contact in response to deceptive behavior [33]: a fine example of how sensitive we are to this form of non-verbal communication.

Some of the most critical factors affecting HRI are the robot's capabilities and performance. Recall that the main requirement for creating a (dis)trust relationship is commitment from the trustee: once that is established, the trustor's positive or negative judgment is determined by their perception of their reliability. If the robot is perceived as reliable, then it will be trusted; otherwise, it will be distrusted. Regardless of other factors coming into play, such as cultural background and the uncanny valley, we will find it hard to place our faith in a robot that doesn't seem capable of carrying out its duties. One robot, in particular, learned this the hard way.

In 2017, the British Broadcasting Corporation (BBC) aired *Six Robots & Us*, a documentary TV series that examined the interactions between humans and social robots. For one of their episodes, the showrunners asked some researchers of Herriot-Watt University of Edinburgh to program a Pepper robot for it to offer customer service inside a grocery store [34]. This robot, nicknamed Fabio, was placed in a shop of the chain Margiotta Food & Wine with the purpose of helping customers by responding to inquiries and offering some occasional small talk. The owners and their staff were excited at the idea of working with a robot and indeed their reactions to its presence were very positive. Unfortunately, their customers did not share their same enthusiasm: in the span of a couple of days, it became clear that Fabio was not able to perform its job efficiently. First of all, background noise in the store would often prevent it from understanding the questions it was asked, forcing the clients to repeat themselves multiple times. What's worse is that even when the robot was able to understand the query, it would not offer particularly helpful advice: when a woman, for example, asked where she could find milk, Fabio replied that it was in the fridge section, but gave no directions on how to reach it nor did it offer to accompany the customer. In an attempt to repurpose it for another job, the robot was tasked to give out food samples, but people ended up avoiding it in favor of one of the human employees. After a week, the owners decided that it was time to pull the plug.

What can we learn from the story of poor Fabio? Despite the initial welcoming reaction of the grocery store staff, both the employees and the shoppers witnessed the robot's incapability to perform its designated tasks effectively. This gradually eroded their trust in it, to the point where they no longer considered it an asset for the job. The owners could perceive its commitment to the task but also deemed it unreliable: a combination of factors that, as we already know by now, leads to distrust. Arguably, perceived capability is one of the most influencing factors in determining the trustworthiness of a robot, often overshadowing the others. One could enter an interaction with a positive bias or a good first impression, just to lose all their trust as soon as they start experiencing the limits and boundaries of technology. This is also the reason why sensationalistic media are detrimental to the advancement of robotics, as they generate unrealistic and often unfair expectations towards what these machines can achieve.

Teaming together

Collaboration between people has been, through history, the key to obtaining the grand achievements of humankind. We are social creatures held together by communal bonds and organized into complex social structures, which provide fulfillment for many needs that range from basic survival to intellectual and emotional expression. This tendency to aggregate, to work as part of groups, and to collaborate with others to achieve common goals is not to be dismissed as a quirk, rather it has been one of the key factors in our success as a species.

This means that, because of evolutionary reasons, we find the concept of teams very attractive: research shows that people need very little incentive to form and feel part of a group. Membership in a team enhances communication, trust, and effort, giving credence to the expression "the sum is greater than the parts." In fact, human teams are so effective (at least, most of the time) that we just can't help but try to get non-humans, such as dogs, involved in them. Unfortunately, these efforts have generally not fully succeeded and in practice, it's far more effective to have a specialized handler work alongside the animal rather than trying to treat the latter as a full-fledged teammate. But what about other kinds of entities, let's say something neither human nor animal?

Historically, computers have been considered unable to be teammates: they are powerful tools, but they lack an appropriate level of communication and coordination. As we have learned, though, robots are embodied agents that can trigger a broad range of social responses and can be designed specifically to support human activities: this means that they have the potential to be peer members of a group. In 2007, scientists from around the world gathered at the International Conference of Human–Robot Interaction, whose theme was "Robot as a Team Member" to explore this possibility. The attendance at the event demonstrated a growing interest in the scientific community to elevate robots from mere tools to peers capable of acting as independent and autonomous partners. Despite this, evidence suggests that we should proceed with caution. A paper from the same year authored by Victoria Groom and Clifford Nass of Stanford University [35], critiqued that, at present, robots can't be good teammates because they lack some of the essential abilities that they should display: first and foremost, the capacity to access the same mental models as humans. A mental model is an explanation of how something works or should be done, inclusive of goals, strategy, and motivations, which helps create a common ground for communication and action. Additionally, Groom and Nass argue that robots set false expectations about themselves and what they can achieve, making collaboration difficult. This is often due to our own tendencies to project social identities onto these artifacts, causing us to expect more than they can deliver. This, inevitably, triggers a negative response when the expectations are

unmet.

Nevertheless, there is an ongoing significant effort to make robots into good teammates. The U.S. military, for example, is investing into the development of autonomous agents that will be able to assist soldiers on the battlefield. Autonomy is important for these artifacts, because otherwise they would not be perceived as peers but rather as tools, similar to telecontrolled robots that are guided inside radioactive areas or collapsing buildings for exploration and rescue purposes.

But why is it so important to make robots into teammates? It's because team composition influences our level of trust. We naturally display different attitudes towards tools versus peers. If we perceive a robot as our teammate, we are more likely to trust it because we believe that we share a common goal and are prepared to protect each other's interests. Being part of a team creates a sense of identity that affects how we perceive our companions. In other words, belonging to a team comes with (trust) benefits.

As insightful as this may be, it is not only team membership that influences our judgment, but also the composition of the team itself. Consider the concept of conformity, which we discussed while examining social influence: the inherent tendency in human nature to align with the answers, choices, and decisions made by the majority of the group. Social scientists have identified two distinct types of conformity: normative and informative. Normative conformity drives individuals to conform to a group to gain acceptance. For example, a high school student may start wearing certain brands of clothing or styles that are popular among their peers to avoid standing out or being ostracized. On the other hand, informative conformity involves the belief that the group possesses competent and accurate information. During a classroom discussion, our high school student might change their opinion on a topic after hearing a convincing argument from a classmate whom they perceive as knowledgeable and well-informed.

In 2022, researchers from Bielefeld University, Germany, attempted to measure the effects of these conformity types when the group consists of humans or robots [36]. They designed an experiment where participants had to select the optimal candidate for a job by looking at some resumes. These contained an equal mixture of positive and negative qualities that made the choice ambiguous, with none of them emerging as the clear best choice. After making their decision, participants were informed of the choices made by other group members and given the opportunity to retain or alter their decision. The experiment was conducted in three conditions: with groups composed entirely of humans or robots, or a mixture of both. At the end of the trial, the subjects were asked some questions aimed at understanding which kind of conformity had driven their choices (such as if they thought that the group knew the answer better than they did).

The findings revealed that normative conformity exerts a stronger influence in interactions with humans, while informative conformity is more pronounced when interacting with robotic teammates. This seems intuitive: we feel less pressured to gain acceptance from artificial agents, and at the same time, there's a common bias toward assuming machines are better at analytical tasks. However, the most interesting discovery of this study is that, when dealing with hybrid teams comprising both humans and robots, both forms of conformity come into play, leading to increased overall conformity among participants. In other words, we tend to adapt more often to the decisions of mixed groups. This can have interesting implications on the way in which we design teams that involve both humans and intelligent agents.

If the robot is not considered a peer, it falls into either a subordinate or a superior role, both carrying different expectations. For instance, in 2019, the American software corporation Oracle conducted a survey about AI in the workplace [37]. Surprisingly, the published results showed that 64% of the interviewees would trust a robot more than their manager. This is a clear indication of how AI is gradually transforming our relationship with technology and society, or perhaps more simply, it's just a sign of how much people don't like their bosses. Unfortunately for them, however, it doesn't seem like robots will be taking over managerial positions very soon: recent research has highlighted the fact that we tend to trust less a robot in an authoritative role than one that we consider a peer [38]. The reasons for this are still unclear, but it might have to do with people questioning the robot's legitimacy as an authority figure. In other words, we generally don't like taking orders from a machine.

This gives us an opportunity to reflect upon what makes human judgment more appealing than pure artificial decision-making. It might involve factors such as empathy, compassion, and phronesis, a term coined by Aristotle, which stands for the type of practical wisdom used to know what is good and what is bad. Pure, cold analytical thinking, typical of the current generation of artificial agents, might not be ideal, especially when our jobs depend on it. Would a robot understand that an employee could really benefit from clocking out half an hour early on a certain day? Would it be able to balance the needs of the company and those of its employees? To solve this kind of problem, scientists and researchers worldwide might wish to widen their perspective on pure, number-crunching artificial intelligence and start considering the potentials of artificial wisdom, with its applications for humane robotics: machines capable of displaying sympathy, generosity, compassion, and forbearing. Maybe at that point, we might be more inclined to better trust robots in authoritative roles. Hopefully, this would send a clear message to many people in positions of power that lack those qualities.

The outcome of our interaction with a robot is also significantly influenced by the context in which it takes place. For instance, the success of the emergency guide robot experiment mentioned earlier was dependent on the fact that the participants genuinely believed they were in an emergency situation and therefore made serious decisions. If any of them had a suspicion of what was really going on, namely that the emergency was only being simulated, then their decisions might have been considerably different. "Sure," they might think, "let's follow the sketchy indications of this robot: it's a game, nothing can happen to me." However, it's not just the conditions of the interaction that matter, as the nature of the task itself can also profoundly impact our decisions.

Remember that Mayer, Davis, and Schoorman's definition of trust highlights the fact that the trustor is accepting and exposing a vulnerability to the trustee. The bigger this vulnerability, the stronger must be the willingness to trust: this means that if the perceived risk is big, I will be more careful in judging the other party. This was brilliantly exemplified in a recent experiment conducted by a group of scientists from the universities of Singapore, Southern California, and Washington about what they called "trust-aware decision-making" [39]. They set up a table on which they displayed several objects, namely three plastic water bottles, a fish can, and a wine glass, then programmed a robot arm to pick these objects, one at a time, and place them on a trolley. In their experiments, they paired this machine with a human participant who would act as a supervisor: when the robot started moving towards a target, they could decide if to allow it or intervene and pick up the object. The scientists noted that, although the robot was perfectly capable of handling all the objects on the table, inexperienced participants were afraid it would drop and break the glass and consequently stopped the machine from handling it. They also noted that the best strategy for an optimal collaboration was for the robot to calibrate its partner's trust: that is, demonstrating its capabilities before challenging the trustor's faith. This was done by making the robot collect the low-risk items first, in order to show evidence of its competency, and only then attempt to pick the delicate item. This is proof that training can help a human gain trust in the system by reducing their initial bias and providing knowledge about a robot's real capabilities. When they cannot anticipate what the robot is going to be doing in a certain situation, their trust is inevitably going to decrease.

To overcome this problem and to become easier to understand for people, robots should explicitly communicate their internal states, such as beliefs, reasoning, and motivations. To achieve this, modern robots often use techniques from the field of explainable artificial intelligence (abbreviated as XAI), which is an approach to creating AI systems that humans can easily understand. AI algorithms can be very complex, often operating in ways that even their developers may not fully comprehend when it comes to how specific decisions or predictions are made. This can make it difficult to trust or make sense of AI systems, which is a pressing matter, especially in fields like healthcare or finance, where incorrect decisions can have serious consequences. This is also important for robots, which, as we have seen, need to be accepted by humans in order to perform their duties. Explainable AI tries to solve this problem by making AI systems more transparent and interpretable: this means designing algorithms in a way that makes it easier for humans to understand how they work, how they arrive at conclusions, and what data they are using. By making AI more explainable, we can increase our trust in these systems, better understand their limitations and potential biases, and ultimately use them more effectively in many different applications. Unfortunately, this is not an easy procedure, and it often forces system developers to have to manage a trade-off between the human interpretability of their model against its accuracy. In other words, making AI more explainable can often come at the expense of their performance.

To trust or not to trust?

This concludes our overview of the main factors that affect trust in HRI. To summarize, we have seen that these can be split into three macro-categories: factors that depend on the human (such as cultural background and personality), characteristics that are intrinsic to the robot (for example, its physical design, the way in which it communicates, and its reliability) and, finally, the environmental context in which the relationship unfolds (team composition and the nature of the task they are involved with). Even if empirical evidence from ongoing studies is suggesting that the robot's performance seems to outweigh the other factors, each of them still plays a role in establishing a positive or negative outcome of the interaction. Hopefully, this brief overview has convinced you of a simple truth: trust isn't simple! In fact, it depends on so many variables, many of which are out of our control. Whether between humans, animals, robots, or a mixture of them, there is no magic equation that can allow us to predict with absolute certainty if two parties will end up trusting, distrusting, or remaining neutral to each other. The best that engineers can do is to keep in mind these factors to maximize the chances of an interaction being positive: for example, building robots that avoid the uncanny valley effect and that come across as reliable by committing the least number of mistakes as realistically possible. Of course, we are still very far away from having perfect machines and anyone living with a voice assistant in their home can certainly vouch for how frustrating it is dealing with "smart" technology nowadays, but keep in mind that we are living in the early stages of life of these new technical marvels. Robots have just started to peek out of research labs and are slowly and timidly starting to take their first steps in our world: a scary and fast-paced one, made of complex untold rules, elaborate languages, social expectations, and little space for errors. If it can be scary and stressful for most of us, think about how difficult it can be for them.

At this point you might be reflecting on what you have learned during this chapter and wondering what the best approach to your future interactions with robots should be. Maybe you have decided that you will adopt the strategy of always trusting them, or maybe, on the contrary, you have settled your mind into distrusting this kind of technology no matter what: if it's not perfect, it's not good enough. Or maybe you are okay with certain types of robots, such as waiters, but are unwilling to take any further risk and refuse for your parcels to be delivered by an autonomous drone. Maybe you disagree with the possibility of a robot taking care of the elderly members of your family, or maybe you are so enthusiastic about robots that you have already ordered yourself an artificial dog to keep you company. Whatever your stance might be, my suggestion is to try and avoid the pitfalls of under-trust and over-trust.

Let's imagine just for a moment that, some years in the future, you will require a surgical operation. Your doctors inform you that this specific procedure has a much higher chance of success with no further complications when performed by their brand-new robotic system, due to the finer precision it can achieve in that particular task compared to a human surgeon. Would you be willing to continue not trusting these machines and put your health (if not your life) at stake?

Consider now another, opposite example. In September 2020, the newspaper The Guardian published an article titled:

"Tesla driver found asleep at wheel of self-driving car doing 150km/h" [40]. Tesla is an American company that designs and manufactures electric and self-driving cars that have become internationally famous for being some of the first robotic vehicles to be commercialized worldwide. Self-driving technology is advanced but far from perfect, as several accidents over the past years have proved, some of which have resulted in fatalities. Given this, it would seem wise to constantly apply a certain level of supervision to the vehicle even when in fully autonomous mode (and in fact, this is what the manufacturer expects from the customer). The car featured in that article was stopped in Canada and its passengers were found asleep, with both front seats reclined, suggesting that they had the precise intention of sleeping on the road. If that wasn't enough, the car was also found traveling above the speed limit for the road it was traversing. Once again: would you be willing to put your well-being at risk for misplacing your trust in a robot (in this case, by over-trusting it)?

Considering this, what is the best approach? Maybe the wiser choice would be to stand in the middle ground, evaluating each situation as it comes, but ultimately, it's not my wish to influence your philosophical point of view. What really matters is that, by understanding the clockwork of trust, how it forms and evolves, you now have the tools to take more conscious decisions about who you trust and distrust in your daily life. This is because what we have learned in this chapter is directly applicable to relationships with other humans: the three categories of factors still apply, with little revisions, to our interactions with other people. In the process of learning more about how to create trustworthy relationships with robots, you might have learned more about human nature and how our mind works. Robots act as mirrors that reveal to us the true nature of humanity by allowing us to reason on ourselves from a third-person perspective. As I have anticipated in Chapter 1, by studying robots we can always learn more about ourselves and, by elevating our understanding of what makes us humans, we can think of ways to make better robots. But even setting humans aside, knowledge of the variables that dictate our attitudes towards robots can help us better interact with them whenever we encounter one.

Our journey in the world of trust does not terminate here: we have explored how our trust towards robots ebbs and flows, but we still have to investigate the reciprocal element of this two-way relationship: specifically, how can robots decide if and when we are worthy of their own, artificial trust.

CHAPTER 3

ROBOTS TRUSTING HUMANS

Trust is a two-way street

Throughout Chapter 2, we have learned about trust: what it is, how it arises from the marriage of commitment and reliance, and the factors that can influence its presence or absence during an interaction with a robotic partner. Most importantly, we have understood that we are facing a complex matter that can't be reduced to a set of equations but rather involves rich facets of psychological, cultural, and educational components.

When I proposed my favorite trust definition, back in Chapter 2, I made an assumption that would make the topic easier to tackle: I asked you to consider the case of trust flowing from one party to the other. In other words, we have considered the existence of one trustor (the human) and one trustee (the robot). This is not an unreasonable assumption, as there are many cases in which trust can be considered a one-way relationship: one example of this is when a person trusts a doctor to perform a medical procedure on them. The patient trusts that the doctor has the necessary knowledge, skills, and experience to carry out the procedure safely and effectively. However, the doctor doesn't need to trust the patient.

Despite this, there are cases in which trust is required from both the involved parties, who then act both as trustees and trustors. Two-way trust is required in situations where both parties rely on each other and need to work collaboratively towards a common goal: for example, the success of a romantic or personal relationship and even teamwork. Students and teachers must trust each other for learning to happen and, at the same time, you can never be friends with someone who doesn't trust you.

This seems natural when thinking about relationships between humans, or maybe even animals, but how do robots come into play in these scenarios? From what we have just said, we should not be able to perform teamwork with a robot if there is no two-way trust between us, but this would imply that the robot would have to decide if to trust me or not. Is this even possible?

As we navigate the fast-paced world of technological progress, it's essential to ask ourselves not only if something is technically possible but also if it is ethical, beneficial, and necessary. The question "is this possible?" shouldn't be the first one we ask ourselves, as it can distract us from the more fundamental inquiry: "should it be done?" By taking the time to reflect on the potential consequences of our actions, we can ensure that we create technology that is not only innovative but also aligned with our values and goals as a society.

The first time I stood in a conference room full of academics and suggested that robots should make trust judgments about humans, the audience seemed perplexed, and understandably so. We tend to view robots as tools and helpers, not autonomous entities capable of making their own decisions. It's no wonder that we may be reluctant to grant them the power to make such assessments. After all, granting the ability to evaluate trustworthiness means allowing a machine to say "no" to us: «I don't trust you, so I won't follow your instructions». Is this how the famous rise of the robots is supposed to begin?

While the idea of robots being able to make trust judgments towards humans may sound dramatic, the reality is much less so. Bidirectional trust is required in situations where mutual dependency and cooperation are necessary to achieve a shared goal. For instance, in a collaborative task between a human and a robot, a mistake from either party could impair their joint efforts. Although robots, like any other machine, are subject to errors and failures, humans are also not immune to mistakes. In such scenarios, it becomes crucial for the artificial agent to anticipate when its partner is likely to fail and work around their weaknesses to ensure that the shared goal is achieved. By doing so, it can play a crucial role in augmenting the human's abilities and enhancing their joint performance.

To better illustrate the concept of bidirectional trust, let's consider some examples. Imagine a scenario where a human and a robot are working together to set up a table. They have collaborated on similar tasks in the past, and the robot has learned from experience that the person is somewhat clumsy and prone to dropping fragile items like glassware. Despite the human's positive commitment towards the task, the robot might anticipate a low reliability and decide to take precautions. To minimize the risk of breakage, it could then suggest taking care of the glassware itself while the human handles the tablecloth and cutlery.

In another scenario, a robot could assist with moving furniture or heavy equipment around a room, knowing that its owner does not possess great physical strength. The machine could take on most of the work while refusing any assistance from the owner, thus ensuring that the task is completed safely and efficiently.

In a medical setting, a surgical robot could assess its own probability of success in a particular procedure and estimate that it is much higher than that of the human doctor. In this case, the robot could offer to switch roles, allocating tasks so that both of them are working to their strengths and maximizing the chances of success for the procedure.

In all of these examples, the bidirectional trust between the human and the robot allows for a more effective collaboration and better outcomes for the shared goal.

Now that we have made sure that enabling these robots with such skills is ethical and beneficial, we can focus on our original question: how can we make them perform these elaborate judgments? We have just seen how complex trust is and how many different factors can influence it: do we really have the knowledge to replicate them in an AI system? Is artificial robotic trust similar or different from natural human trust? To answer these questions, we need to introduce a relatively new scientific discipline that will give us the tools to tackle this fascinating challenge.

Sciences of the mind

Humans have marveled at the complexities of the mind for millennia. From the philosophers of ancient Greece to the neuroscientists of today, we have been on a never-ending quest to uncover the secrets of cognition and consciousness. This is important because the mind plays a central role in our existence: it is tied to our sense of self-identity, and it is responsible for perceiving and interpreting the world around us. Understanding how the mind works is a pathway towards understanding who we really are.

Before the scientific revolution, which took place during the 16th and 17th centuries, people studied the mind through philosophical and religious inquiry, rather than scientific methods. Philosophers, theologians, and mystics explored questions about the nature of the mind, consciousness, and the self. We have already discussed some of these ideas in Chapter 1, when we discussed the mind–body problem and several philosophical stances that inspired the design of early artificial intelligences.

In the modern era, the discipline of psychology emerged in the late 19th century as a scientific field of study. Wilhelm Wundt, often considered the founder of modern psychology, established the first laboratory for psychological research at the University of Leipzig in Germany in 1879. Wundt and his students conducted experiments to study the basic elements of human consciousness, such as perception, sensation, and attention. These scientists would rely on introspection: a method of self-reflection that consisted in asking the subjects of their studies to observe and report their mental processes, such as their thoughts and feelings, in response to specific stimuli. This method was of course controversial because it is unreliable and highly subjective.

In the early 20th century, another psychological approach became dominant in the United States: behaviorism. Behaviorists believed that observable behavior was the only valid subject matter for psychological study, and that the mind was a black box that should be ignored. They focused on the relationship between environmental stimuli and the resulting actions, and believed that all behavior could be explained through learning principles. The most famous example is of course "Pavlovian conditioning," named after its inventor: the Russian physiologist Ivan Pavlov, who conditioned some dogs to salivate at the sound of a bell by repeatedly pairing the sound of the instrument with the presentation of food. After several repetitions, the dogs began to salivate at the sound of the bell alone, even when no food was present.

As you can imagine, deciding to ignore the contents of the mind and only focusing on the reaction/stimuli pairing is a coarse approximation of reality that disregards the complexities of our mental life. For this reason, the 1950s saw the birth of the cognitive revolution in which researchers became increasingly interested in understanding what goes on inside that black box. The advent of new technologies, such as computers, led them to realize that the mind could be seen as a device that could represent and process information: an idea that would consolidate in the "mind-as-a-computer" metaphor. According to the latter, the brain is like a computer, with inputs coming from the senses, processing occurring in the brain, and outputs resulting in behavior. This idea led to a period of rapid growth and advancement in the field of psychology and resulted in the foundation of cognitive science: an interdisciplinary field of study that aims to understand the nature of the human mind by bringing together insights from psychology, neuroscience, philosophy, linguistics, computer science, and anthropology.

Before we dive further into this subject, let's try to understand what is meant by "cognition." The American Psychological Association (APA) defines it as the set of mental processes that are involved in learning, remembering, and using knowledge. It's an umbrella term that encompasses several mental skills, such as: attention (the ability to focus on specific information while ignoring other), perception, language, memory, and reasoning. Cognition is a fundamental part of human psychology, and it allows us to act as autonomous entities in the world by learning from experience, adapting to new situations, communicating with others, and solving problems.

The digital age provides scientists with an invaluable tool: by viewing the mind as an information processor, they can write computer simulations that mimic different cognitive processes to test their hypothesis on how the mind works. These simulations, also known as computational models, are based on theoretical knowledge and refined through empirical evidence collected through experiments.

Let's see an example. In 1968, Richard Atkinson and Richard Shiffrin proposed a computational model of memory: the faculty to store, retain, and retrieve information over time. This is a very important cognitive process because it scaffolds the ability to learn: without memory, we would not be able to acquire and recall knowledge. The Atkinson–Shiffrin memory model [41], shown diagrammatically in Figure 3.1, asserts that human memory is divided into three categories: sensory (where information from the senses is stored for milliseconds after its experience), short-term (also known as working memory, where information is stored while other cognitive processes operate on it), and long-term (where information can be stored even for a lifetime). This model describes not only the components that characterize human memory but also the way in which information flows between them: in other words, it offers a complete description of the process, which can be simulated in a computer program and tested through experiments. The Atkinson–Shiffrin model was one of the first to describe memory and fails to capture the complete essence of the process, but it's simple enough to provide an excellent example that should give you a sense of what these models look like.

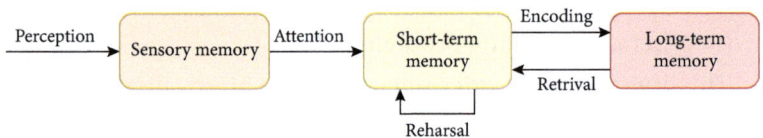

Figure 3.1 The Atkinson–Shiffrin memory model.

When a computational model describes the interlocking work of several cognitive processes, then it is referred to as a cognitive architecture. If the mind is like a complex machine with many different parts that work together to produce our thoughts and behaviors, a cognitive architecture can be imagined as a map or blueprint of this machine, showing how all the different parts fit together and work in harmony.

Cognitive architectures are essential not only in the study of natural systems but also in the design of artificial ones.

An artificial cognitive system is an agent programmed to work autonomously in everyday environments, anticipating outcomes and adapting to an ever-changing environment. Robots that use cognitive architectures are called cognitive robots, and the field that combines AI, cognitive and biological sciences, and robotics is called cognitive robotics.

Cognitive robotics allows us to design robots that possess human-like mental abilities. By bridging robotics with psychology and cognitive science, it provides a framework to study how robots can learn, reason, and interact with humans in a more natural and intuitive way. This makes cognitive robotics an exciting field with vast potential to revolutionize the way we interact with machines. The main difference with traditional AI is that the latter is not particularly interested in solving problems in the same way a human would: in other words, it does not necessarily care about creating biologically plausible models. On the contrary, cognitive robotics cares about the path as much as the destination and aims to create human-like capabilities using human-like solutions.

But why is cognition useful in robotics? There are several reasons, the first of which is that it provides the potential for autonomy. Remember that these machines are meant to operate in the real world, which is a messy, dynamic, and often unpredictable environment. Whether they must survive the rugged surface of Mars or the intricacies of human social lives, cognition provides them with a tool to try and anticipate both the need and outcome of action, other than assisting them in adapting to changing circumstances. An autonomous car, for example, might use a cognitive architecture to adapt to changing driving conditions, such as road construction or heavy traffic.

The second reason is that every system, whether natural or artificial, experiences sensorimotor latency. This is the time difference that occurs in our brain between the acquisition of sensory information and the generation of a physical response to it. For example, if someone throws a ball at you, the light from the event must first reach your eyes, be detected, encoded, and transferred via the optical nerve to your brain. The brain must then process this information, choose an appropriate response, and send signals to the muscles to initiate movement. By this point, the ball has already moved further along its trajectory: if you wait until the ball is within reach to start moving your arm, you will likely miss it. This means that we cannot base our immediate responses on the present alone: rather, we must be able to anticipate the future and act accordingly. Cognition provides us with the ability to predict the ball's trajectory and begin moving our arm and hand so that we are ready to intercept it.

Cognition provides a third advantage to robotics by enabling effective interaction with humans. Cognitive robots can deduce the goals and intentions of the people they are interacting with and behave in a helpful manner. Moreover, humans tend to engage better with other cognitive agents, meaning that their relationships with cognitive robots are likely to be of better quality.

Let's revisit the primary issue that led us on this journey: modeling trust capabilities in robots. As we are attempting to emulate a human cognitive skill in an intelligent machine, it is clear by now that cognitive robotics provides a strong foundation, incorporating a methodology that spans across multiple disciplines. Let's explore the forefront of science and learn how researchers are currently empowering robots to make their own trust decisions when interacting with people in their vicinity.

Artificial trust

The concept of trust is not new in computer science. In the context of cybersecurity, a system is deemed trustworthy if it is capable of behaving in a secure and reliable manner, by protecting sensitive data or securing communication channels. For example, let's say you're shopping online and come across a website that you've never heard of before. You might be hesitant to enter your credit card information because you're not sure if the website is trustworthy. However, if you see a trust seal or badge on the website issued by a certified authority, your confidence in the website's security and legitimacy may increase.

The kind of trust we are interested in has a slightly different connotation. As we have previously discussed in the context of human–robot interaction, considering trust as a one-way relationship would be an oversimplification: not only do humans need to be willing to trust robots for them to achieve their full potential, but robots may also need to evaluate the humans they interact with. As an example, a robot acting as a hospital assistant may not accept instructions from someone it believes to be untrustworthy, such as a person pretending to be a doctor to access sensitive medical information.

Although trust is a widely researched topic in the field of robotics, most of the state-of-the-art literature focuses on the trust that flows from the human to the robot, neglecting the cases in which the robot is the trustor of the relationship.

This largely unexplored area has only recently begun to emerge as a field of interest and stands on the frontier of human knowledge. Scientists including myself have started naming this kind of trust "artificial trust." The latter involves all the forms of cognition and intelligence required for a robot to decide whether to trust another entity, be it a human or a fellow robot. It involves the ability to make decisions and take actions based, for example, on an assessment of the reliability, competence, and honesty of the other party, without relying on human intervention.

Artificial trust is essential for the safe and effective operation of autonomous systems, as it allows them to make informed decisions and take actions based on their autonomous evaluation of the entities they interact with. But how can robots be programmed to use this synthetic form of trust?

Using the cognitive robotics approach, a good starting point would be to identify which human mental faculties are connected to the ability to evaluate someone's trustworthiness. By taking inspiration from the way nature works, it may be possible to replicate the process in a computer system. The scientific literature in human psychology is a rich source of knowledge on how biological minds solve this problem, and one possible approach to pursuing this challenge comes from the field of developmental psychology. As you may recall, the latter is the branch of science that deals with how people change and grow from infancy to adulthood, both physically and mentally. In 2011, a group of developmental psychologists from the University of California authored a paper titled *The Development of Distrust* [24]. The scientists wanted to test at what age pre-school children can assess the reliability of a source of information, and in particular when they are able to distinguish between a helpful and deceptive assistant. Their hypothesis was that this ability is deeply connected with Theory of Mind: the ability to understand that others have their own thoughts, beliefs, desires, and perspectives that may differ from one's own. We have already encountered this concept in Chapter 2, when talking about the effects of age on the human's ability to trust. To determine if I can trust someone, I first need to verify how well their beliefs, goals, and intentions align with mine.

The experiment described in this paper consisted in a sticker-finding game: a child subject was placed in front of two boxes and had to find a sticker hidden inside one of them. Before choosing which one to open, they received a suggestion from an adult, who was secretly acting either as a helper or a tricker: the former would suggest the correct location, while the latter would try to mislead them by pointing to the wrong box. The subject would then have to decide if to accept or reject the advice, based on how trustworthy they would consider their advisor. This experimental setting is summarized in Figure 3.2.

At first, the child did not play the game themselves but instead observed a few pre-recorded rounds through a video. The goal of this was to familiarize them with the adult informants and their habits. Only after that were they invited to participate in person in a game session, where they could apply their own decision-making.

The results of this experiment show that children aged 3 and 4 were unable to distinguish between helpers and trickers and would, in most cases, trust their advice. Five-year-olds, on the other hand, demonstrated the ability to trust the

Figure 3.2 The Vanderbilt experiment on the development of distrust. A child searches for a sticker located in one of two boxes and requires the assistance of an adult.

helpers and distrust the trickers by more consistently reject-ing the advice received from the latter. Is it just a coinci-dence that this is also the age when Theory of Mind begins to mature, providing them with a tool to better evaluate other people's behavior?

Scientists often remind us that "correlation does not imply causation." This phrase is used in statistics to explain that just because two events occur together, it doesn't necessarily mean that one causes the other. For instance, just because someone always carries an umbrella when it starts raining, it doesn't mean that carrying the umbrella causes the rain. Other factors, such as the season, location, or time of day, are likely involved. Therefore, when analyzing data, it's crucial to avoid assuming that one thing causes another merely because they are some-how correlated. One of the most amusing examples of this is provided by Tyler Vigen on his website *Spurious Correlations:*[1] the number of films in which Nicolas Cage appeared between 1999 and 2009 correlates remarkably well with the number of people who drowned in a pool in the U.S. during the same period. The curves nearly overlap perfectly, yet no one would seriously claim that there is any causal relationship between the two!

To ensure that the relationship between the ability to trust and the maturity of Theory of Mind was causal rather than merely correlational, the research team used a set of questionnaires that assessed the children's understanding of the interactions they had just experienced. The goal was to grade the child on the Theory of Mind Scale, a psychological measure used to evaluate an individual's ability to attribute mental states to oneself and others. By using this tool, the scientists were able to confirm that there was indeed a correlation between the ability to assess the trustworthiness of others and the development of Theory of Mind.

This discovery can guide our inquiry into artificial trust: if we want robots to make their own trust judgments, a promising approach would be to equip them with a form of synthetic Theory of Mind. Why not, then, replicate the sticker-finding experiment we discussed, but this time replace the human child with a robot equipped with an appropriate cognitive architecture? If we can design and program this robotic brain to perform similarly to the human subjects in the original experiment, it would indicate that we are on the right path, as the robot would essentially

---

[1] https://www.tylervigen.com/spurious-correlations.

replicate the behavior of humans in that context. In other words, we would have succeeded in designing an artificial intelligence that mirrors certain human cognitive processes.

In the early days of my doctoral studies, I was directly involved in this research. My colleague, Massimiliano Patacchiola, who was then working at the University of Plymouth, developed a computational model based on probability theory that could simulate the reasoning of the children participating in the sticker-finding experiment [42]. I was tasked with embedding this model into a cognitive architecture designed to replicate a set of cognitive abilities such as perception, reasoning, and memory, and implementing it on a Pepper robot [43]. Following the script of the original experiment by Vanderbilt et al., the robot was given the opportunity to familiarize itself with the informants and their behavioral patterns before playing the game (Figure 3.3). When the game commenced, the robot had the ability to make its own decisions, accepting or rejecting the humans' advice based on its assessment of their trustworthiness (Figure 3.4). Additionally, the robot was able to develop its own personality, which determined how trusting it would be towards unfamiliar people, mimicking the "trust versus mistrust" stage of human development theorized by Erikson (as discussed in Chapter 2). If the robot encountered more helpers than trickers, it developed a more trusting personality and found it easier to trust strangers. Conversely, if its experience was dominated by people trying to hinder its progress in the game, it would become more distrustful of new individuals it encountered. Finally, Pepper had the opportunity to test its meta-reasoning abilities by predicting the suggestions that each human would most likely give based on its knowledge of them (Figure 3.5).

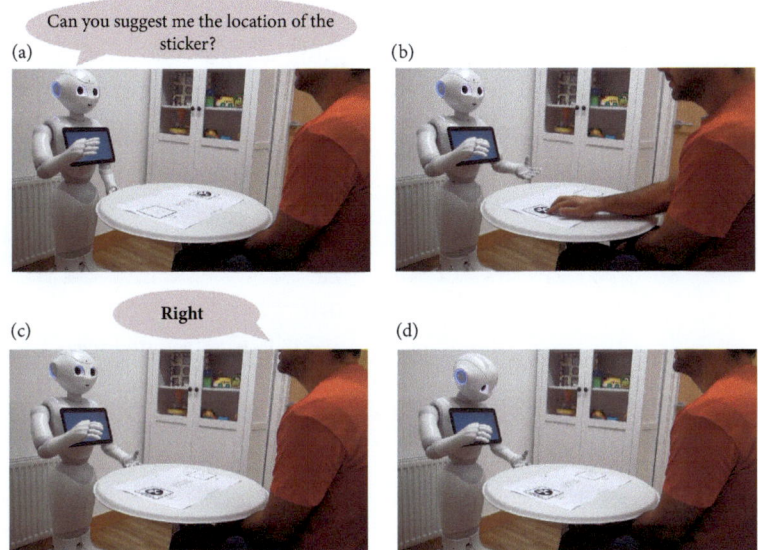

Figure 3.3 Pepper familiarizes with the informant (in this case a tricker) by asking them to give it advice on the position of the sticker on the table. This would happen several times for each human.

This research project proved to be significant because it was one of the first to shift the perspective on trust in human–robot interactions from the viewpoint of the human to that of the robot. In the following years, scientific literature gradually began to feature more papers discussing and experimenting with artificial trust. For instance, while our work focused on the holistic trustworthiness of humans, a group of scholars from the University of Michigan, U.S., developed a computational model that evaluates artificial trust in relation to individual tasks by considering both the human's capability and the task's difficulty [44].

For a robot to develop artificial trust towards a human, it must understand which of the human's features correlate with

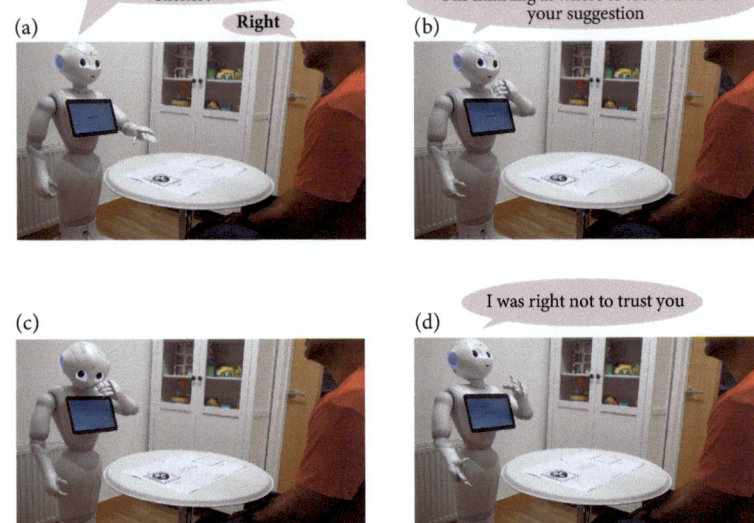

Figure 3.4 Once familiarized with the tricker, Pepper wisely decides to reject their misleading suggestions.

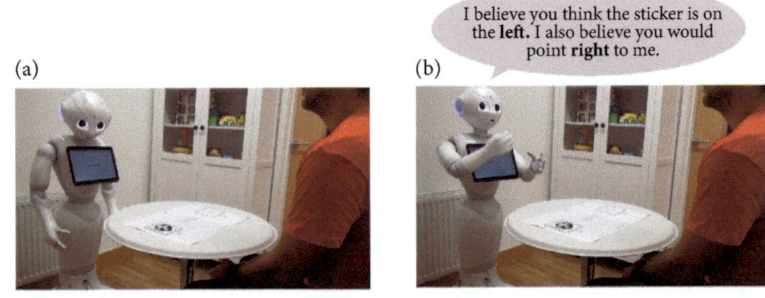

Figure 3.5 Pepper could also answer some meta-reasoning questions, anticipating the suggestion of each specific informant. In this case, it correctly predicted the incoming deception by the tricker.

their trustworthiness and how these can be observed through their behavior. Is it simply a matter of ability, where a lack of skill makes the person unreliable and therefore untrustworthy? Or must the robot assess their actions to gauge benevolence or integrity, thus evaluating their commitment to the task? Researchers from the Institute of Cognitive Sciences and Technologies of the Italian National Council of Research (ISTC-CNR) propose that, when estimating trust, an agent should consider whether the other party can perform an action (ability), will perform that action (disposition), and whether any external factors might favor or hinder the achievement of the objective (opportunities). They developed a symbolic AI system capable of reflecting on these three dimensions and providing a numerical trustworthiness score for a set of professionals across different tasks [45].

When artificial trust is not directed towards a human trustee, it can manifest in two other forms that are equally important in the design of a fully autonomous cognitive agent. One such form is inter-robot trust, which refers to the degree to which one robot trusts another in a collaborative task or activity. Just as humans need to trust each other in team settings, artificial agents must also establish trust among themselves to work together effectively. This type of trust is particularly crucial in multi-robot systems, where robots need to coordinate their actions to achieve a common goal by communicating and sharing information with each other. For example, Rishwaraj and Ponnambalam from Monash University Malaysia conducted an experiment where three robots were tasked with exploring an area by scanning cells of different colors scattered throughout the environment. One robot was designated as the leader, but unknown to the other robots, it had a fault in its sensors that occasionally caused it to misinterpret the colors it detected. After working together for some time, navigating the area, scanning, and reporting the colors to the rest of the team, the two follower robots realized that the information transmitted by their leader was incorrect. They then began to distrust it and negotiated a new leadership between themselves [46].

Self-trust, on the other hand, refers to a robot's ability to assess its own abilities and limitations and make decisions based on that self-evaluation. This type of trust is crucial for agents operating autonomously or in environments where they have limited or no direct supervision. By continuously monitoring its status and performance, a robot can determine whether it can perform a certain task. For instance, a robot aware of its poor manual dexterity, perhaps from repeated failures in handling objects, might choose not to attempt grabbing a delicate item, distrusting its own capabilities in that context.

My colleagues Marta Romeo, Francesco Semeraro, and I have identified a growing trend in the number of scientific papers published on this topic between 2013 and 2023, with an increasing number of researchers interested in exploring the computational aspects of trust. Some of these papers focus directly on artificial trust, while others address the related, though distinct, issue of enabling robots to estimate the level of trust that humans have in them. This is a valuable skill for a machine, as it allows it to adapt its behavior based on how trustworthy its human partners perceive it to be. We've already seen an example of this: in Chapter 2, we described how a robot could pick and place non-frail items before fragile ones to demonstrate its capabilities to a user and calibrate their trust in it [39]. We have decided to refer to this form of trust as "natural trust," to distinguish it from artificial trust, and we have termed the mathematical and algorithmic techniques aimed at addressing either as computational models of trust.

This field of study is relatively new, and as a result, we currently lack elegant theories or paradigms to fully explain the concept of artificial trust in robots. However, as our understanding of this topic evolves, we will eventually be able to formalize it into structured frameworks, similar to the three-factor model of human–robot trust that guided the discussion in Chapter 2. At this stage, many questions remain unanswered, including the similarities and differences between artificial and biological trust. For instance, we have observed that, for humans, a robot's performance during an interaction often outweighs other factors influencing trust. Should this also be the case for mechanical trust? Can a robot enter an interaction with biases, much like how humans account for personality and culture?

This uncharted territory can be both exciting and intimidating. It is daunting to explore the edge of human knowledge, where no minds have ventured before. However, what has always set humans apart is the spirit of exploration. The realm of artificial trust invites us to venture into new frontiers and uncover its secrets, some of which are certain to change our world forever.

Consider the potential impact of artificial trust on society: autonomous cars capable of switching to self-driving mode if they detect that the driver is unfit to drive, whether due to intoxication or fatigue. Robot educators that could adapt their teaching style and content to the unique learning styles and needs of each student, offering a more effective and personalized experience. Social robots that could assist us around the house, autonomously deciding that children shouldn't handle knives in the kitchen. Robot assistants for the elderly that could determine whether to trust their patient's promise to take their medications. The possibilities are endless, limited only by our imagination.

CHAPTER 4

TRUST IN THE FUTURE

Moral machines and ethical quandaries

Many people are rightfully concerned about the uncontrolled rate of improvement of artificial intelligence-based systems and the way they will affect society. As we saw at the beginning of this book, Ray Kurzweil's Law of Accelerating Returns means that technological progress is happening much faster now than ever before. My grandfather was born in a world without television, computers, and mobile phones, and by the end of his life, the Internet, smartphones, and social media were commercially available to anyone. A few years later, AI stopped being just science fiction. What is going to happen five years into the future? Will the society that lies 20 years ahead of us have any kind of resemblance to the one we know today?

People are anxious because they are uncertain about the future. Not knowing what lies ahead can be exciting for some, but stressful for many. Adding to this concern is the fact that AI development is currently in the hands of corporations that have repeatedly demonstrated a priority for capital income over the best interests of humankind and planet Earth, by "moving fast and breaking things."

This brings us to a crucial discussion about ethics in AI, as it becomes increasingly important to ensure that these powerful technologies are developed and used in ways that are aligned with the values and well-being of society. Ethics can be seen as the philosophy that explores the boundaries between moral and immoral behavior, seeking to determine what is good and what is bad. It's a discipline that raises complex questions, many of which don't have straightforward answers. However, these questions have to be asked to ensure we are moving in the right direction.

For the purposes of our discussion about intelligent robotics, we can distinguish two fields of ethics that we should be interested in. The first one is artificial intelligence ethics, which David Leslie defines as: "a set of values, principles, and techniques that employ widely accepted standards of right and wrong to guide moral conduct in the development and use of AI technologies" [47]. This branch of ethics tries to formulate standards and good practices for the correct use of this technology, with the purpose of avoiding misuse that might lead to discrimination and violation of individual rights.

To fully understand what this means, let's consider a relatively recent example. Google's Photos app can automatically attach labels to the pictures uploaded by its users. These labels are meant to make it easier to search for specific topics or events. For instance, one might search for all their pictures involving the beach or birthday parties. In 2015, Google found itself at the center of a significant scandal when web developer Jacky Alciné pointed out that the app was classifying pictures of his Black friends as "gorillas." The episode went viral, and rightfully so: despite the machine not explicitly trying to be racist (or having any intentions at all, for the matter), what happened was completely unacceptable. Google swiftly extended its apologies and promised to investigate a solution. As of the time of writing this book, nearly a decade later, it seems that Google's solution was to restrict the app from labeling anything as a monkey, rather than addressing the problem at its root.

This is a classic case of bias in artificial intelligence, which occurs because modern systems are trained on datasets provided by human developers. If the data reflect existing societal biases, such as those related to gender, race, or socioeconomic status, the model can learn and perpetuate these biases. For example, in August 2024 I asked ChatGPT to generate a set of images of CEOs without specifying any other details. Out of the ten pictures it created, all depicted men. When I asked the chatbot to generate images of a successful person, only one out of ten portrayed a woman. This kind of bias is harmful because it echoes and amplifies societal issues that we are striving to overcome. We don't want our robots to treat people differently based on the color of their skin or the clothes they are wearing.

Trust can indeed be seen as a kind of bias, especially in the context of how it influences decision-making and judgments. We have already seen how factors such as personal cultural background and reciprocal group affiliation can significantly influence our propensity to trust someone. Yamagishi and Yamagishi [48] have described the propensity (or lack thereof) to trust as a form of cognitive bias, which we might use when assessing a new acquaintance. They refer to this as "general trust" and contrast it with "knowledge-based trust," which develops later in a relationship through experience and observation. The latter is the kind of trust that I would have towards my partner, which leads me to believe that they are not willing to cheat on me.

The fact that trust relationships begin with implicit bias means we must exercise great caution when designing software that interacts with us. When creating systems that will be at the receiving end of our trust, we must ensure they are not designed in a way that could deceive us about their true capabilities. Social robots, for example, are often designed to mimic human behaviors through carefully engineered appearance, facial expressions, or gestures, which can create a false sense of empathy and understanding. Users might believe the robot genuinely understands and cares about them, leading to emotional dependency and misplaced trust. Moreover, these robots can be programmed to manipulate users' emotions or behaviors.

The fourth rule of the U.K. Engineering and Physical Sciences Research Council's (EPSRC) Principles of Robotics states that "robots are manufactured artefacts: the illusion of emotions and intent should not be used to exploit vulnerable users" [49]. This opens a complex discussion regarding the ethical use of deception in social robotics. By this point in the book, you might be reasonably convinced that social robots are inherently deceptive. For instance, they might simulate emotions and display them toward the user solely to enable natural and multimodal interaction.

Consider the case of Paro, a seal-like robot used for elderly companionship (Figure 4.1). While this machine does not intentionally disguise itself as a real seal, some authors criticize that the illusion of sentience and cognition it may evoke in some people could be considered a form of deception [50]. However, this type of deception is not always negative. Some argue that if deception enables a robot to function as a social entity, then it might be a necessary skill for its integration into our society. In my view, we should establish a clear distinction between beneficial and harmful deception, defined by the impact on individuals and society. A robot pretending to have emotions to facilitate clearer communication and increase its acceptance may be ethical, as long as it does not use those artificial emotions to manipulate people. On the other hand, a robot that pretends to be more capable than it is could have negative repercussions, particularly for the youngest and oldest members of our society.

Figure 4.1 Paro, a seal-like robot designed by Takanori Shibata Aaron Biggs, Flickr user ehjayb, CC BY-SA 2.0, via Wikimedia Commons.

The second field of ethics that we are interested in takes the opposite perspective. Machine ethics refers to the science of enabling an intelligent machine to be guided by and reason upon ethical rules when deciding how to act and respond to a given situation. While AI ethics is more focused on what we, as humans, can do to develop and train more ethical systems, machine ethics is concerned with how to empower agents to make morally sound decisions.

James Moor made a significant contribution to this field when, in 2006, he published an influential paper in which he defined a hierarchy of machine ethical agency [51]. At the lowest level are "ethical impact agents," which can be evaluated for the ethical consequences of their actions. In other words, at the lower end of the spectrum we find systems that are not capable of reasoning about ethical dilemmas, but nonetheless have an ethical impact on society. Moor provides a brilliant example of such an agent. In Qatar, camel racing is a popular activity among the wealthy. Because camel jockeys need to be light to avoid hindering the animal, young boys were often abducted from poorer regions and forced to ride. In response to international criticisms from commercial partners in the West, in 2004 Qatar developed robot jockeys that could ride racing camels. The ethical impact of this machine was the prevention of large-scale slavery across the nation.

The second level of the hierarchy is "implicit ethical agents," which are designed to avoid unethical outcomes. These agents are not meant to reason about ethics but are instead designed to avoid certain negative outcomes. For example, consider an autonomous vehicle programmed to avoid collisions with pedestrians. The machine does not reason about the ethical implications of hitting someone; rather, it is simply programmed to do its best to avoid that outcome.

The third level of the hierarchy is "explicit ethical agents," which are machines that can finally reason upon ethics. An explicit ethical agent can assess a given situation, decide on a course of action, and evaluate its outcomes in light of the ethical rules it has been provided with to determine if the outcomes are acceptable. For instance, consider a robot designed to assist elderly individuals and tasked with making decisions about their care. This machine might have to decide when to override a patient's request for a risky activity by considering principles of beneficence (doing good) and autonomy (respecting the patient's wishes). Researchers are currently studying the best ways to implement explicit ethical agents. Bremner et al. [52] argue that these systems need to be proactive, meaning they should not merely evaluate the moral soundness of a certain course of action but also be able to generate alternative solutions that would prevent any violations of their ethical guidelines. They demonstrate their ideas by implementing a system that can generate plans, ethically evaluate their outcomes, and select the one that seems to produce the best overall result. This architecture is used within a robot tasked with preventing another robot from walking into danger. The ethical robot evaluates the path its partner is taking, predicts potential dangers, and, where necessary, formulates a trajectory for itself to intercept and stop the other robot without, if possible, putting itself at risk.

Finally, the highest level of the hierarchy is defined by Moor as "full ethical agents." These are hypothetical machines that can make explicit moral judgments and justify them, much like a human with consciousness, intentionality, and free will would do. To create artifacts that fit this level of ethical agency, we would first need them to fully understand what they are doing, which is perhaps the greatest limitation of today's AI systems (remember the Chinese room experiment described in Chapter 1).

At the time of writing, the goal for researchers in this domain is to create well-rounded explicitly ethical agents. But what does this have to do with the topic of trust? As it turns out, trust can be an invaluable tool when trying to act ethically. Imagine a healthcare robot working in a hospital or care home. One day, someone approaches it and pretends to be a doctor to gain access to sensitive information about one of its patients. Interactions with robots are intended to be natural, and there are not many situations where a robot will ask someone to authenticate themselves through a password or other security measures we commonly use with computers. A naive robot might simply assume that this person has legitimate access to the data. However, a robot capable of performing trust assessments and evaluating them might refuse the request. Of course, this is a double-edged sword, as, like humans, robots cannot always be certain that they will trust the trustworthy and distrust the untrustworthy. This is why AI systems should be designed to minimize biases.

A term that has been increasingly used to describe this kind of behavior is intelligent disobedience. Originally coined to describe instances where a service animal, such as a guide dog, disobeys its owner's instructions to make a better decision, the term has recently been applied to robots as well. An intelligently disobedient robot might use a combination of trust estimation capabilities and ethical reasoning to recognize when it is receiving commands that might violate ethical or safety rules, and then refuse to comply, or even better, propose alternative solutions. For example, an autonomous car that refuses to speed through a red traffic light, prioritizing the law and safety over the command, is a valid instance of intelligent disobedience in action. Briggs and Scheutz [53] demonstrated this concept by placing a robot on a table, ready to receive instructions from a human operator. The agent would obey commands to sit or stand, but when the person asked it to walk forward, the robot objected, saying, "Sorry, I cannot do this," and justifying its refusal by pointing out the lack of support ahead, which would cause it to fall and get damaged. The robot only agreed to move forward when the human promised to catch it. However, I wonder, what would happen if the human failed to keep their promise? This is where I argue that trust could come into play: the robot could use its artificial trust mechanisms to assess whether to trust the human to catch it, ultimately deciding whether to obey or disobey the command. In other words, trust capabilities can make ethical decision-making more nuanced.

If properly developed, intelligent disobedience could become a powerful tool, enabling future robots to perform their duties more effectively. However, it requires the machine to possess a good understanding of the environment in which it operates, other than a certain degree of artificial wisdom.

The main challenge is that ethics is not an exact science. It is a complex branch of philosophy that often raises more questions than it answers. This is not a shortcoming of the discipline, but rather a reflection of the intricate nature of what is being investigated. Moral conduct is not universal; it can vary significantly between cultures and even among individuals within the same social group. In this sense, artificial ethics shares with artificial intelligence the same challenge: attempting to formalize and implement in a machine an aspect of humanity that we do not yet fully understand. As a result, we approach the problem by approximating what we know, hoping to uncover the underlying mechanisms of human nature. Perhaps the virtuous cycle of robotics, which draws from scientific knowledge with the aim of giving back more than it takes, also applies to ethics. After all, as Mark Coeckelbergh brilliantly states in his book *Robot Ethics* [54]: "asking about robots is also always asking about humans along with their morality, practices, and institutions. [...] Robots are mirrors that show us the often-beautiful yet also darker sides of humans as well as their moral thinking and doings."

A society of trust

Human–robot interactions, much like the ones between people, do not occur exclusively in a dyadic manner, meaning between just two agents at a time. In Chapter 2, we have only begun to explore relationships that involve larger groups, such as teams, but our perspective can be broadened even further to consider the societal implications of integrating intelligent machines into our daily lives. While social robots are not yet widespread enough to consistently capture the attention of international agencies and policymakers, there has been a growing awareness of the ethical, legal, and social implications of AI. As this technology becomes more deeply embedded in our everyday experiences, governments and international organizations have begun to recognize the need for regulatory frameworks that ensure these technologies are developed and deployed responsibly.

The European Union's AI Act is the world's first comprehensive legislative effort in this domain. Initially proposed by the European Commission in 2021, it came into force in August 2024. The AI Act is a European Union regulation, like the famous General Data Protection Regulation (GDPR), meaning it is a binding legal act across all member states of the Union and automatically becomes part of their national law. Its purpose is to regulate the use of AI not designed specifically for military, national security, research, or non-professional uses by classifying it into four levels of risk.

The lowest level is "minimal risk," where most of today's AI systems, such as video game agents or spam filters, are expected to fall. Due to their low likelihood of posing any risk to the rights or safety of EU citizens, these applications will not require additional regulations, and everyone will be free to use them. One level up are "limited risk" applications, such as chatbots or systems that can generate or manipulate multimedia (e.g., deepfake creators). These applications will be subject to transparency obligations to ensure that people are aware they are interacting with a machine or with an artifact that was not produced by humans.

Next, we have "high-risk" applications, which can significantly impact people's health, safety, or rights. Examples include software systems capable of rejecting job candidates based on their CVs or those used in education and healthcare. In my opinion, most social robotics applications will fall into this category. These products will need to be evaluated and approved before being made available to the public.

Finally, at the top of the list are "unacceptable risk" systems, which are outright banned due to the ethical issues they pose. Examples include social scoring software, machines that manipulate people's behavior, and facial recognition systems in public spaces. Additionally, the AI Act includes a category for "general-purpose artificial intelligence," such as ChatGPT, which will be subject to transparency obligations.

Although this book primarily focuses on the interpersonal aspects of trust between humans and robots, it is crucial to acknowledge that these policy efforts play a significant role in shaping the broader environment in which these interactions occur. The AI Act appears to be a promising tool for potentially fostering public trust in these technologies. The fact that an international governmental agency has declared a particular robot fit for public use could influence personal biases (what we defined as human-related factors in Chapter 2) and alter how people relate to these intelligent machines.

What the European Union is striving to achieve is the emergence of trustworthy AI, a concept that has gained significant traction as part of these regulatory efforts. It is not just about making systems that function correctly, but mostly about ensuring that these systems align with societal values and can be trusted by users to act in their best interests. This concept promotes several key principles, including fairness, transparency, accountability, and respect for human rights. However, building trustworthy agents presents several challenges. For example, ensuring transparency in complex machine learning models is not always straightforward. As a result, many existing models make decisions that are not understandable by humans. Imagine you are a doctor who must decide whether to send a patient to surgery because your robot assistant has diagnosed a pathology with 80% confidence but cannot explain its reasoning or how it reached that conclusion. Would you trust it? How would responsibility be attributed if something went wrong? Researchers worldwide are putting considerable effort into solving this problem, but the issue remains unresolved. As robots become more autonomous, the need for mechanisms that ensure their alignment with human values becomes increasingly important.

One way to foster the trustworthiness of these systems is by ensuring meaningful human control. This concept was initially proposed to address responsibility gaps in autonomous weapons (for example, who is responsible for a death caused by an autonomous robot soldier or drone?) but has since evolved to encompass a broader range of systems beyond military applications. The core idea is that humans should always maintain control and retain moral responsibility over the actions of an autonomous machine. For instance, a robot should not be able to hire or fire someone without the approval of a human supervisor but should instead limit itself to providing suggestions. Similarly, a medical robot should not prescribe any treatments to patients without the green light from a human doctor.

Meaningful human control and trust are deeply interconnected. When humans are in charge, they can ensure that robots operate as expected, align with ethical norms, and remain safe against misuse. This reassurance can help users perceive the system as reliable, which, as we have learned, is one of the conditions required for trust to form. Meaningful human control also introduces accountability: if something goes wrong, there is a person who can be held accountable and take corrective action, which can prevent the erosion of trust in technology.

Finally, having a human supervisor in control can also mitigate the fear of autonomy that many people in the West experience.

In conclusion, while significant strides have been made in establishing policy frameworks like the AI Act, much work remains to be done. The rapid pace of technological advancement often outstrips the development of corresponding regulations, creating gaps between technology and law. To bridge these gaps, it is essential for policymakers, industry experts, and researchers to collaborate closely to create an environment where human-robot interactions are grounded in trust, supported by robust legal and ethical foundations.

The journey so far

We have embarked on a journey into the world of robotics and AI, with a particular focus on social robots: intelligent and autonomous machines designed to interact with us in our daily lives. While this technology might seem futuristic, we should not underestimate the rapid pace of progress. What was considered unthinkable just a few years ago is now a reality.

In a not-so-distant future where we will be sharing our lives with these intelligent artifacts, it is crucial that our relationships with them meet the highest standards. From this perspective, we must begin considering the implications of trust when interacting with these devices, particularly how it forms and what factors influence our attitudes toward them.

We have seen that trust forms when a trustor perceives both commitment and reliability in a potential trustee. Conversely, the presence of commitment but an unreliable attitude quickly leads to distrust. Of course, it is also possible to have neither: if the other party shows no commitment toward us, forming any judgment about them would be irrelevant. This principle applies not only in our interactions with humans but also in our relationships with robots. In the case of robots, their commitment is often easy to deduce from the context: for example, I cannot expect my robotic vacuum cleaner to help me book a medical appointment, but I can reasonably expect it to clean the floors of my apartment. Whether it does so reliably or not will determine whether I trust it to do its job without supervision.

Performance is likely the factor that influences us the most: all other things being equal, it would be difficult to have faith in a robot that does not behave according to our expectations. However, many other factors can affect our trust. Some of these are related to us, the human entering the relationship. Our personality, history, and cultural background set our initial biases toward these machines, and our trust can also depend on the state in which we experience the encounter, such as the emotions that we are feeling. Other factors depend on the robot itself. While we've already discussed performance, we shouldn't overlook the importance of physical appearance. After all, we do judge a book by its cover: it's a psychological mechanism deeply ingrained in us through millennia of evolution.

Finally, our trust can also depend on the context in which we interact with the robot, including the reciprocal roles we assume and the specifics of the task at hand. Ultimately, each person will weigh these factors differently, and there is no magic formula to predict in advance whether someone will develop feelings of trust, distrust, or neutrality toward a robot. However, being aware of these factors can help us approach these situations with greater awareness, alerting us to the otherwise unconscious reasons driving our decisions. This, in turn, can enhance our understanding of how we interact with others, contributing to the virtuous cycle in robotics where the knowledge we gain about robots deepens our understanding of the mechanisms that make us human.

On the other side of the coin, it would be a mistake to overlook the potential that artificial trust can bring to a relationship. If we aim to develop truly autonomous collaborative robots, we must begin embedding these machines with mechanisms, such as cognitive architectures, that enable them to perform their own trust evaluations. This capability allows robots to compensate for our mistakes, anticipate potential failures, and ensure that our shared goals are achieved despite any obstacles that may arise. Recent research in cognitive robotics is paving the way for this, drawing direct inspiration from the workings of the human mind.

Despite all this progress, we are still far from fully understanding what happens in both natural and artificial minds. Currently, the only way to study the trust mechanisms between a person and a robot is through experiments in controlled settings, which limit the extent and duration of the interaction in both time and space. While the discoveries made so far are certainly representative of the inner workings of the human brain, they still fall short of capturing the depth of a long-term relationship that could develop between the two.

If anthropomorphism has taught us anything, it is that we will likely form strong bonds with these artificial agents, potentially even considering them part of the family. In such an intimate context, how do trust dynamics change, and how do the processes of earning, breaking, and repairing trust compare to those of a casual interaction that lacks the weight of shared history? We might assume that these dynamics would mirror those in human relationships, but until we can rigorously test this through scientific methods, this remains only an assumption.

On the other hand, our artificial trust models are still quite limited in scope and are not yet able to capture the full richness of human trust. While we may successfully design cognitive architectures that process information in a way that elicits behaviors resembling the ability to assess trustworthiness, we have yet to paint the complete picture. For instance, how would artificial emotions interfere with the trust estimation process? If they are so important for humans, why should they be less valuable for robots? This, of course, leads to many other intriguing and currently unresolved dilemmas, such as whether robots should and can possess emotions. But as you can see, much more research is needed to expand our understanding.

Alan Turing once wondered whether machines could think, and today it is clear that they can. The difference lies in how they think compared to us. Exploring these complex questions may be the task of the next generation of scientists, engineers, philosophers, and researchers.

By studying robotics and AI, we learn a great deal about ourselves as humans. The advent of computers led to the establishment of cognitive science, providing us with a completely new and revolutionary way to study the mind. This, in turn, has enabled us to create even more advanced technology. Today, AI is closer than ever: it has already been around for years in the form of subtle systems that often operate behind the scenes, not capturing much public attention. One notable example is recommender systems: a type of AI that predicts a user's preferences or interests and suggests items, products, or content they are likely to enjoy. These systems are widely used in e-commerce, social media, and entertainment platforms to enhance user experience and increase engagement. For instance, Netflix and YouTube can suggest streaming content you might like, Amazon can recommend products you are more likely to purchase, and Spotify can create dynamic playlists tailored to your personal taste in music.

Today, we are witnessing a significant and noticeable shift in how we think about and interact with technology. The rise of powerful and easily accessible artificial intelligence systems, such as ChatGPT, is beginning to have a profound impact on our society and the way we think about it. In his book *The Most Human Human* [7], Brian Christian reflects on how the autofill feature of popular email providers has influenced our writing style, encouraging us to choose shorter or simpler words instead of more complex or creative language. The latter is commonly considered the cornerstone of humankind, the faculty that truly makes us special. If technology is gaining the power to alter the way in which we use it, and learns how to use it itself, what does that say about our humanity and our special place in the universe?

Until now, we've known that robots exist, but they haven't really touched our daily lives. They've mostly been confined to factories assembling parts or to university research laboratories. However, this is about to change: commercial robots are already starting to enter the market, and it likely won't be long before we see them more frequently in our everyday environments. Before we know it, they'll be everywhere, making our lives easier while simultaneously challenging our understanding of who we are and what it means to be human. When that time comes, we must be ready to answer some fundamental questions that will shape the course of our future. By learning how we interact with these extraordinary machines, we can collectively steer humanity in the best possible direction.

The road ahead

As we approach the end of this book, take a moment to reflect on everything you have learned. Whether you are an expert in computer science, AI, or robotics, or simply a curious reader with no prior knowledge of the subject, I hope you have gained valuable insights that will accompany you in your future endeavors.

At this point, I invite you to consider some open questions designed to encourage your personal introspection. There are no right or wrong answers to these questions; what truly matters is that you take the time to engage with them and reflect on how your own experiences and perspectives shape your attitudes toward robots.

1. How does your personal cultural or societal background influence your trust in others (humans or robots)?
2. Are there certain stereotypes or narratives portrayed by popular culture and media that shape, perhaps unconsciously, your perception of robots?
3. How comfortable would you be with the idea of trusting a robot to perform a task that is important to you, such as caring for a loved one or driving you to a destination?

4. What kind of characteristics would a robot need to have for you to trust it? Would it need to have emotions, empathy, or a sense of humor?

5. How might robots challenge our traditional notions of trust towards other humans?

In conclusion, for those interested in exploring the intricate connections between humans, robots, and AI further, I would like to recommend some additional reading materials. The following bibliography offers a comprehensive list of related works, including books, scientific publications, and online resources that we have encountered throughout our discussion. These resources delve deeper into the topics covered in this text, with some focusing on the history and philosophy of AI, while others provide a more scientific perspective on cognitive science and robotics. Regardless of your interests or background, I believe these resources can offer valuable insights into the fascinating world of machine intelligence and its impact on our lives.

REFERENCES

[1] P. Robinette, W. Li, R. Allen, A. M. Howard, and A. R. Wagner, 'Overtrust of robots in emergency evacuation scenarios', *2016 11th ACM/IEEE International Conference on Human-Robot Interaction (HRI)*, IEEE, 2016, pp. 101–108.

[2] R. Kurzweil, *The Age of Spiritual Machines: When Computers Exceed Human Intelligence*, Penguin Publishing Group, 1999.

[3] K. Capek, *RUR (Rossum's Universal Robots)*, Penguin, 2004.

[4] G. A. Lindeboom, 'Descartes and medicine', in *Descartes and Medicine*, Brill, 1979.

[5] A. M. Turing, 'Computing machinery and intelligence', in *Parsing the Turing test*, Springer, 2009, pp. 23–65.

[6] S. Russell and P. Norvig, *Artificial Intelligence: A Modern Approach*, in Always Learning, Pearson, 2016.

[7] B. Christian, *The Most Human Human: What Talking with Computers Teaches Us About What It Means to Be Alive*, Doubleday, 2011.

[8] C. Choi *et al.*, '12 events that will change everything', *Scientific American*, Jun. 01, 2010. Accessed: Sep. 23, 2022. [Online]. Available: https://www.scientific american.com/article/12-events-that-will-change-everything/.

[9] V. Braitenberg, *Vehicles: Experiments in Synthetic Psychology*, in Bradford Books, MIT Press, 1986.

[10] Neil Harbisson, 'I listen to color', TEDGlobal 2012. Accessed: Sep. 21, 2022. [Online]. Available: https://www.ted.com/talks/neil_harbisson_i_listen_to_color/.

[11] P. Loviken, N. Hemion, A. Laflaquiere, M. Spranger, and A. Cangelosi, 'Online learning of body orientation control on a humanoid robot using finite element goal babbling', *2018 IEEE/RSJ International Conference on Intelligent Robots and Systems (IROS)*, IEEE, 2018, pp. 4091–4098.

[12] S. K. Card, T. P. Moran, and A. Newell, *The Psychology of Human-Computer Interaction*, CRC Press, 1983. doi: 10.1201/9780203736166.

[13] F. Teixeira, 'The Worst Volume Control UI in the World'. Accessed: Sep. 30, 2022. [Online]. Available: https://uxdesign.cc/the-worst-volume-control-ui-in-the-world-60713dc86950.

[14] Boston Dynamics, 'Introducing Spot Classic (previously Spot)'. Accessed: Oct. 03, 2022. [Online]. Available: https://www.youtube.com/watch?v=M8YjvHYbZ9w.

[15] R. C. Mayer, J. H. Davis, and F. D. Schoorman, 'An integrative model of organizational trust', *Academy of Management Review*, vol. 20, no. 3, pp. 709–734, 1995.

[16] K. Hawley, *Trust: A Very Short Introduction*, in Very Short Introductions. OUP Oxford, 2012.

[17] P. F. Strawson, *Freedom and Resentment and Other Essays*, Routledge, 2008.

[18] S. Shpall, 'Moral and rational commitment', *Philos Phenomenol Res*, vol. 88, no. 1, pp. 146–172, 2014.

[19] R. Hakli and P. Mäkelä, 'Moral responsibility of robots and hybrid agents', *Monist*, vol. 102, no. 2, pp. 259–275, 2019.

[20] S. Marchesi, D. De Tommaso, J. Perez-Osorio, and A. Wykowska, 'Belief in sharing the same phenomenological experience increases the likelihood of adopting the intentional stance toward a humanoid robot', *Technology, Mind, and Behavior*, vol. 3, 2022, p. 11.

[21] D. C. Dennett, 'Mechanism and responsibility', in *Essays on Freedom of Action*, T. Honderich, Ed., Routledge, 2015, pp. 157–184.

[22] K. E. Schaefer, 'Measuring trust in human robot interactions: Development of the "trust perception scale-HRI"', in *Robust Intelligence and Trust in Autonomous Systems*, R. Mittu, D. Sofge, A. Wagner, W. F. Lawless, Eds., Springer, 2016, pp. 191–218.

[23] S. Baron-Cohen, A. M. Leslie, and U. Frith, 'Does the autistic child have a "theory of mind"?', *Cognition*, vol. 21, no. 1, pp. 37–46, Oct. 1985, doi: 10.1016/0010-0277(85)90022-8.

[24] K. E. Vanderbilt, D. Liu, and G. D. Heyman, 'The development of distrust', *Child Dev*, vol. 82, no. 5, pp. 1372–1380, Sep. 2011, doi: 10.1111/J.1467-8624.2011.01629.X.

[25] J. D. Henry, L. H. Phillips, T. Ruffman, and P. E. Bailey, 'A meta-analytic review of age differences in theory of mind', *Psychol Aging*, vol. 28, no. 3, pp. 826–839, Sep. 2013, doi: 10.1037/A0030677.

[26] A. L. Vollmer, R. Read, D. Trippas, and T. Belpaeme, 'Children conform, adults resist: A robot group induced peer pressure on normative social conformity', *Sci Robot*, vol. 3, no. 21, p. 7111, Aug. 2018, doi: 10.1126/SCIROBOTICS.AAT7111/SUPPL_FILE/AAT7111_SM.PDF.

[27] S. A. Midson, *Cyborg Theology: Humans, Technology and God*, Bloomsbury Publishing, 2017.

[28] T. Ramge, *Who's Afraid of AI?: Fear and Promise in the Age of Thinking Machines*, The Experiment, 2019.

[29] D. Gallimore, J. B. Lyons, T. Vo, S. Mahoney, and K. T. Wynne, 'Trusting robocop: Gender-based effects on trust of an autonomous robot', *Front Psychol*, vol. 10, no. MAR, p. 482, 2019, doi: 10.3389/FPSYG.2019.00482/BIBTEX.

[30] M. Mori, K. F. MacDorman, and N. Kageki, 'The uncanny valley', *IEEE Robot Autom Mag*, vol. 19, no. 2, pp. 98–100, 2012, doi: 10.1109/MRA.2012.2192811.

[31] N. Friedman, K. Love, R. A. Y. Lc, J. E. Sabin, G. Hoffman, and W. Ju, 'What robots need from clothing', *DIS 2021 – Proceedings of the 2021 ACM Designing Interactive Systems Conference: Nowhere and Everywhere*, pp. 1345–1355, Jun. 2021, doi: 10.1145/3461778.3462045.

[32] I. Torre and L. White, 'Trust in vocal human–robot interaction: Implications for robot voice design', in *Voice Attractiveness: Studies on Sexy, Likable,*

and *Charismatic Speakers*, B. Weiss, J. Trouvain, M. Barkat-Defradas, and J. J. Ohala, Eds., Springer, Singapore, 2021, pp. 299–316. doi: 10.1007/978-981-15-6627-1_16.

[33] E. Schellen, F. Bossi, and A. Wykowska, 'Robot gaze behavior affects honesty in human-robot interaction', *Front Artif Intell*, vol. 4, p. 51, May 2021, doi: 10.3389/FRAI.2021.663190/BIBTEX.

[34] T. Mogg, 'Pepper the robot fired from grocery store for not being up to the job'. Accessed: Jan. 19, 2023. [Online]. Available: https://www.digitaltrends.com/cool-tech/pepper-robot-grocery-store/.

[35] V. Groom and C. Nass, 'Can robots be teammates?: Benchmarks in human–robot teams', *Interact Stud*, vol. 8, no. 3, pp. 483–500, Jan. 2007, doi: 10.1075/IS.8.3.10GRO.

[36] L. Masjutin, J. K. Laing, and G. W. Maier, 'Why do we follow robots? An experimental investigation of conformity with robot, human, and hybrid majorities', *ACM/IEEE International Conference on Human-Robot Interaction*, vol. 2022, pp. 139–146, Mar. 2022, doi: 10.1109/HRI53351.2022.9889675.

[37] 'AI@Work Global Study'. Accessed: Jan. 26, 2023. [Online]. Available: https://go.oracle.com/LP=86149?elqCampaignId=230263.

[38] S. P. Saunderson and G. Nejat, 'Persuasive robots should avoid authority: The effects of formal and real authority on persuasion in human-robot interaction', *Sci Robot*, vol. 6, no. 58, p. 5186, Sep. 2021, doi: 10.1126/SCIROBOTICS.ABD5186/SUPPL_FILE/SCIROBOTICS.ABD5186_MOVIE_S1.ZIP.

[39] M. Chen, S. Nikolaidis, H. Soh, D. Hsu, and S. Srinivasa, 'Trust-aware decision making for human-robot collaboration: Model learning and planning', *ACM Transactions on Human-Robot Interaction (THRI)*, vol. 9, no. 2, pp. 1–23, 2020.

[40] L. Cecco, 'Tesla driver found asleep at wheel of self-driving car doing 150 km/h', *The Guardian*. Accessed: Jan. 26, 2023. [Online]. Available: https://www.theguardian.com/world/2020/sep/17/canada-tesla-driver-alberta-highway-speeding.

[41] R. C. Atkinson and R. M. Shiffrin, 'Human memory: A proposed system and its control processes', *Psychology of Learning and Motivation – Advances in Research and Theory*, vol. 2, no. C, pp. 89–195, Jan. 1968, doi: 10.1016/S0079-7421(08)60422-3.

[42] M. Patacchiola and A. Cangelosi, 'A developmental Bayesian model of trust in artificial cognitive systems', *2016 Joint IEEE International Conference on Development and Learning and Epigenetic Robotics, ICDL-EpiRob 2016*, pp. 117–123, Feb. 2017, doi: 10.1109/DEVLRN.2016.7846801.

[43] S. Vinanzi, M. Patacchiola, A. Chella, and A. Cangelosi, 'Would a robot trust you? Developmental robotics model of trust and theory of mind', *Philosophical Transactions of the Royal Society B*, vol. 374, no. 1771, p. 20180032, 2019.

[44] H. Azevedo-Sa, X. J. Yang, L. P. Robert, and D. M. Tilbury, 'A unified bi-directional model for natural and artificial trust in human-robot

collaboration', *IEEE Robot Autom Lett*, vol. 6, no. 3, pp. 5913–5920, Jul. 2021, doi: 10.1109/LRA.2021.3088082.

[45] R. Falcone, M. Piunti, M. Venanzi, and C. Castelfranchi, 'From manifesta to krypta: The relevance of categories for trusting others', *ACM Transactions on Intelligent Systems and Technology (TIST)*, vol. 4, no. 2, Apr. 2013, doi: 10.1145/2438653.2438662.

[46] G. Rishwaraj and S. G. Ponnambalam, 'Integrated trust-based control system for multirobot systems: development and experimentation in real environment', *Expert Syst Appl*, vol. 86, pp. 177–189, 2017.

[47] D. Leslie, 'Understanding artificial intelligence ethics and safety: A guide for the responsible design and implementation of AI systems in the public sector', *The Alan Turing Institute*, Jun. 2019, doi: 10.5281/ZENODO.3240529.

[48] T. Yamagishi and M. Yamagishi, 'Trust and commitment in the United States and Japan', *Motiv Emot*, vol. 18, no. 2, pp. 129–166, Jun. 1994, doi: 10.1007/BF02249397/METRICS.

[49] M. Boden *et al.*, 'Principles of robotics: Regulating robots in the real world', *Conn Sci*, vol. 29, no. 2, pp. 124–129, Apr. 2017, doi: 10.1080/09540091.2016.1271400.

[50] A. Sharkey and N. Sharkey, 'We need to talk about deception in social robotics!', *Ethics Inf Technol*, vol. 23, no. 3, pp. 309–316, Sep. 2021, doi: 10.1007/S10676-020-09573-9/METRICS.

[51] J. H. Moor, 'The nature, importance, and difficulty of machine ethics', *IEEE Intell Syst*, vol. 21, no. 4, pp. 18–21, Jul. 2006, doi: 10.1109/MIS.2006.80.

[52] P. Bremner, L. A. Dennis, M. Fisher, and A. F. Winfield, 'On proactive, transparent, and verifiable ethical reasoning for robots', *Proceedings of the IEEE*, vol. 107, no. 3, pp. 541–561, 2019, doi: 10.1109/JPROC.2019.2898267.

[53] G. M. Briggs and M. Scheutz, '"Sorry, I can't do that": Developing mechanisms to appropriately reject directives in human-robot interactions', in *2015 AAAI Fall Symposium Series*, 2015.

[54] M. Coeckelbergh, *Robot Ethics*, The MIT Press, 2022. doi: 10.7551/mitpress/14436.001.0001.

INDEX